CONTENTS

QUANTITATIVE METHODS

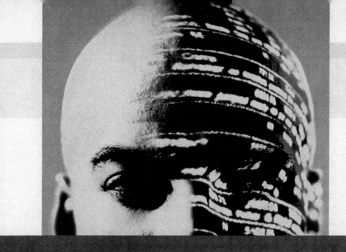

QUANTITATIVE METHODS

A SHORT COURSE

Jon Curwin and Roger Slater

Sophia Kittoch

CENGAGE
Learning™

For product information and technology assistance, contact **emea.info@cengage.com**.

For permission to use material from this text or product, and for permission queries, email **clsuk.permissions@cengage.com**.

British Library Cataloguing-in-Publication Data
A catalogue record for this book is available from the British Library.

ISBN: 978-1-84480-905-9

Cengage Learning EMEA
Cheriton House, North Way, Andover, Hampshire, SP10 5BE, United Kingdom

Cengage Learning products are represented in Canada by Nelson Education Ltd.

For your lifelong learning solutions, visit
www.cengage.co.uk

Purchase your next print book, e-book or e-chapter at
www.CengageBrain.com

Printed by Lightning Source, UK

PREFACE

This book has been written to help you make sense of the maths and statistics you are likely to be taught on a short introductory course. Most certificate, diploma and degree programmes will include some numeracy. Working effectively with numbers will be important whether you are a student of business, engineering or nursing. We hope this book will help you in a number of ways. It is intended to be concise, help you when you need help, reflect current practice and provide web support.

This is not a big book. It does seem that at a time when many courses are delivered in one semester, the books are getting bigger. It is easy for authors to feel they need to include everything. We have purposely chosen to be selective. The 14 chapters reflect what we think will typically be included in a one-semester course. The introduction is a guide to the book, the importance of the personal computer and the supportive web site. Rather than make the book bigger, we have included some topics that might be of interest on the web. The book includes a revision of basic sums. Even if your course does not include a review of the mathematics you might need, we do suggest you look at this. The book does include working with data in a variety of ways and, introduces important concepts like the value of money over time and probability.

This book should help you when you need help. We assume little previous knowledge and most chapters can be read independently of the others. It would be nice to think that all the readers would read a chapter a week and even try all the exercises at the end of each chapter. The chapters should be self-explanatory. If you miss a few weeks or did not fully understand a topic at the time, working through a chapter should help you catch up. In addition to exercises at the end of each chapter, we give fully worked answers. You do need to know what your course expects of you. We suggest you check the information given in the course outline. Most courses will give details of the outcomes expected. You should also check whether past examination papers are available from your library. The final chapter gives advice on preparing for any coursework or examinations.

This book should give you the numeracy skills that courses and employers can reasonably expect. We are now in an age of information. In reality there is no shortage of data. The available data may or may not have the required validity or may or may not answer your particular questions but the chances are you will find plenty of it. This book should help you be selective with the data and help you decide whether additional data is required. You will also develop your communication skills. There is a difference between what the data is and what the data means. Quantitative methods is not just about number calculation but also about the interpretation. Work with numbers is typically computer-based.

Spreadsheets are used extensively and the book gives a number of examples of the type of printout you can expect.

The book has a supportive web site. We have chosen to limit the size of the book. You may find the book complete (not everyone likes the web!) or you may seek extra information. The web site has been designed to help you. As a student, you can find the PowerPoint presentations that would be used for typical lectures (no substitute for attending lectures), additional exercises with answers and some additional topics that may be covered by your course but not included in the book, for example, on networks.

We would also welcome feedback from you. We would like to think that this book will help you with your course now and will also make you more effective with numbers in the future. We can even add to the web site if you would like.

Jon Curwin
Roger Slater

August 2003

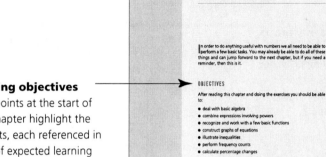

Learning objectives
Bullet points at the start of each chapter highlight the concepts, each referenced in terms of expected learning outcomes.

Conclusions
Each chapter has a summary section which links to the learning outcomes and suggests practical applications of the chapter material.

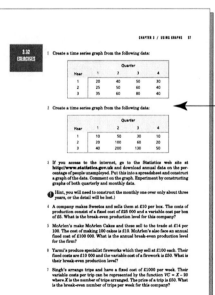

Exercises
Exercises are provided to check understanding of the concepts and give practice in calculation.

GUIDED TOUR

Glossary terms
Basic definitions and
clarification of key
terms are given in a
glossary at the end of
the book.

Answers to exercises
Within each chapter there are detailed annotated
answers to give you immediate feedback on every
question.

Web sites
References within the text to commercial and
governmental web sites provide deeper
knowledge about certain topics.

Web reference
There is a companion
web site for the book
that includes additional
exercises and
annotated answers,
additional topics and
PowerPoint slides.

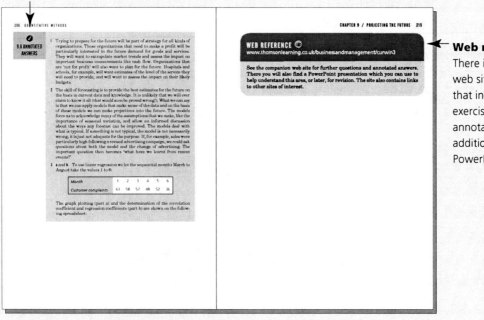

ONLINE TEACHING AND LEARNING RESOURCES

Visit the companion web site for *Quantitative Methods: A Short Course* at **www.thomsonlearning.co.uk/businessandmanagment/curwin3** to find valuable further teaching and learning material including

FOR STUDENTS

- Additional exercises designed to test your understanding and improve your results
- Fully worked solutions to additional exercises to help you arrive at the right answer
- Additional material to help you read around the subject and further your knowledge
- PowerPoint slides for use as an overview to each chapter and as a revision aid
- Web links to relevant sites mentioned in the text
- Online Glossary to explain key terms
- Excel supplement including a Guide to Excel and additional Excel exercises
- Interactive multiple choice questions to test your understanding of the chapter

FOR LECTURERS

- A password-protected site with teaching material
- Additional material, also found in the student section, to allow you to explore the subject in greater depth with your students
- Additional exercises to use with your students in the classroom or as part of their home study
- Fully worked solutions to all additional exercises
- Additional PowerPoint slides of key material in the book

Chapter 1

INTRODUCTION

Welcome to this new book. We hope this book will become a useful companion as you begin your studies of quantitative methods. We will be working with numbers in a variety of ways. Essentially we want to develop those skills referred to as *numeracy skills*. Numbers should make sense and you should be able to make sense of numbers! Working with numbers is more than just getting right answers. An awareness of magnitude, expressed in numerical terms, allows more effective description, more effective communication and develops the ability to ask the right questions.

There are very few problems with real content that cannot benefit from some numerical description and analysis. We soon begin to ask questions like 'how many', 'how long' and 'how much'. We only have to consider some of the issues of the workplace, like sales promotion or training, to recognize the importance of numbers. Those involved in sales will want to know about past performance and likely trends. This book considers ways to clarify and summarize past information. This book also considers ways of using collected data to make projections about the future. Those involved in training will soon want to know 'how many' for 'how long'. Numbers are there to help us (really!) and we want to consider ways of getting more value from numbers.

OBJECTIVES

After reading through this book and working through the exercises you should be able to:

● work with numbers with increasing confidence

● identify and create sources of numerical information

● summarize and present data in a variety of ways

● be aware of a range of mathematical and statistical techniques

● perform a range of calculations

● understand the importance of statistical or mathematical modelling

● use spreadsheets when appropriate

1.1 | WHAT CAN YOU EXPECT?

Will this be a difficult book? No. This book is designed to consolidate and build on the knowledge you have. We accept that you may have forgotten much of the mathematics you did at school. A revision of basic mathematics is offered in Chapter 2. Few assumptions are made about what you remember. Concepts from mathematics are explained as needed. We also show how useful mathematics (and statistics) can become when applied to problem situations. This book is written to support an introductory course in quantitative methods (often referred to as maths and stats, or as QM). The book is particularly suitable for a one-semester course in statistics at an introductory level.

Will you want to read this book? Yes. The book works through a series of topic areas using chapters of roughly equal size. A typical introductory semester course would cover one of these topic areas each week. By reading a chapter a week and working through some of the examples, you should develop the topic competence. If you miss a week (or two), then you can always catch up with additional reading. The chapters are designed to be fairly independent, so that you can jump around the chapters and expect them to make sense. You will find examples of the calculations needed with additional exercises at the end of each chapter. We have also included the worked solutions to these exercises at the end of each chapter, given as *Annotated answers*, so that you can immediately check your progress. The importance of your computer is acknowledged and you will see many solutions presented as spreadsheet output.

The initial focus is on the revision and development of your mathematical skills. Even if you feel confident with your sums we would still recommend that you work through Chapter 2 (it's still nice to know that you know). Chapter 3 then considers a few important applications of basic mathematics and graphical representation. It is not likely that the simple use of mathematics will solve real problems directly, but it will allow you to think more clearly about these problems and the issues involved. Consideration of how a company can break even or the 'best' solution to a specified production problem does begin the process of business modelling. If we can model a problem, perhaps a significant business problem, we can think about what is important, what has been missed, what has been assumed, what the likely options are and what are likely to be the consequences of possible actions.

Chapters 4, 5, 6 and 7 consider the various ways you can present and represent numerical information. It will be expected that we can graphically represent information (Chapter 4), calculate descriptive statistics like the mean and standard deviation (Chapters 5 and 6) and make percentage comparisons (Chapter 7). We know that effective communication is an important personal skill – a business competence. In many situations we would like the figures to do the talking for us. We are sure that you would agree the benefits of working with factually based information. It is therefore important to be able to share this information with others. In addition to description, we need to explore relationships. We want to know how one factor, perhaps advertising, affects another, perhaps sales. Chapter 8 looks at the relationships between variables and Chapter 9 looks at movements over time.

You will again see the use of basic mathematics in Chapter 10 when we consider the value of money over time. The use of algebra will allow us to address a question like 'what will this future sum of money be worth to me now'. In Chapter 11, we move away from the idea of certainty (that we know for sure what some future sum of money will be) to the ideas of uncertainty, probability and risk. Chapters 12 and 13 look at sources of information. At times you will need to generate your own data (Chapter 12) and at other times you will use data that has been collected for other purposes (Chapter 13). In many cases you will need both types of data. Finally, Chapter 14 reviews the type of assessment you might have on your course and gives you advice on how you might prepare for this assessment.

This book should help you with any one-semester or short introductory course in quantitative methods. It should also provide you with a future source of reference. Whatever your future career, you are likely to work with numbers again. This book should inform your management of any data and enhance what you are able to say, perhaps supported by the use of Excel and PowerPoint.

1.2   THE USE OF COMPUTERS

The use of computers should make our life easier. In practice, quantitative methods should not require a lot of manual computation. A quality calculator or a package like Excel should be doing the work for you. But you do need to understand the answers. We show, with a number of examples, how the calculations work. You should also try these and get a feel for the data. It is worth developing an intuition for the right answer. You should know from the experience of the data that the average journey time is, say, 3 hours rather than 30 hours or 0.003 hours. Given the importance of spreadsheets a number of the answers are shown in Excel format. We encourage you to develop your spreadsheet skills and many questions assume that you can do this. We find that many courses combine QM with IT skills, so you may be learning about spreadsheets at the same time as you are learning about QM. The final section of the book and a part of the web site has a guide to the use of Excel.

1.3 | A TEACHING NOTE

This book is not intended to replace class contact, although we know that contact time with students may be limited. Typically a course may expect the introduction to quantitative methods to be given over a one-semester period on the basis of one lecture each week and one supportive seminar. This book is intended to help students make best use of the class contact they have. The topics covered assume little previous knowledge and are typical of a short course. Students can 'dip in' to these chapters before or after the lecture. They should find illustrative examples in the text and detailed answers given to the exercises at the end of each chapter. Students can be directed to parts of the book if they report subject difficulty. We have also included supportive material on the web for staff and students.

1.4 | THE WEB SITE

Whilst the book is free standing and can be used on its own, we have also constructed a web site to support and supplement the book. In a book of this kind it is not possible to include all the topics that an introductory course might require. The web site includes the coverage of topics that may be of specific interest to certain courses. If required, notes and exercises are available to staff and students on *networks* and on *confidence intervals*. We have covered the most commonly found topics for these introductory QM courses in the text but intend to add further topics to the web site in response to user feedback. You can reach the web site at **www.thomsonlearning.co.uk/businessandmanagement/ curwin3**

The site also includes extra questions and annotated answers to allow you more practice on the various topics. As with the book, you need to attempt the questions before reading through the answers, but QM is a topic where practice is one of the key elements in successfully passing the module or unit.

Also on the web site are a series of 12 PowerPoint presentations, one for each of the main chapters. Students might want to watch these as a way of getting an overview of a topic, or as a revision aid. Lecturers are at liberty to use these slides themselves, if they so wish, within their own lecture series.

We do not expect the web site to be static. We intend to add materials from time to time, especially in response to feedback. We also invite others to contribute to the web site, maybe by submitting case studies which can be included, with suitable acknowledgement, or links to their own sites to illustrate other topic areas.

**The web site address is www.thomsonlearning.co.uk/
businessandmanagement/curwin3**

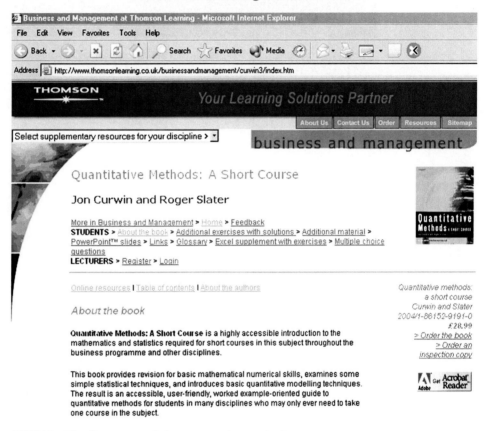

FIGURE 1.1　The first page of the companion web site

1.5　FURTHER READING

This book is a starting point for you to become familiar with some quantitative techniques. We have deliberately not included theoretical derivations, nor much background material, since there is rarely time to cover such material in single semester, introductory courses. However, if you wish to find more information and details of other quantitative methods, you could read *Quantitative Methods For Business Decisions*, Jon Curwin and Roger Slater, 5th ed. (Thomson Learning, 2002). This book contains web site links to illustrate and develop the topics, but these and further links are also on the companion web site. Web addresses tend to move and change over time, but some notes on Searching the Web written by Mark M. H. Goode of Cardiff Business School are available on our web site. You will also find suggestions and help on searching in the text in Chapter 13.

WEB REFERENCE 👁
www.thomsonlearning.co.uk/businessandmanagement/curwin3

See the companion web site for further questions and annotated answers. There you will also find a PowerPoint presentation which you can use to help understand this area, or later, for revision. The site also contains links to other sites of interest.

Chapter 2

BASIC SUMS

In order to do anything useful with numbers we all need to be able to perform a few basic tasks. You may already be able to do all of these things and can jump forward to the next chapter, but if you need a reminder, then this is it.

OBJECTIVES

After reading this chapter and doing the exercises you should be able to:

- deal with basic algebra
- combine expressions involving powers
- recognize and work with a few basic functions
- construct graphs of equations
- illustrate inequalities
- perform frequency counts
- calculate percentage changes

2.1 | WHERE DO YOU WANT TO START?

Numbers are a part of everyday life and being able to add, subtract, multiply and divide is essential. You may just want to know how much you will be charged for three cinema tickets which cost £4.50 each, or you might work behind a bar and need to add up the cost of a round for fifteen people who each ordered a different drink. You could even be faced by the task of advising someone on investing their inheritance of £2.5m.

Whilst all of this is familiar, some students we have met cannot remember the order in which you do these things. Just to remind you, there is a mnemonic, known as **BEDMAS**, to help:

Brackets
Exponentiation
Division
Multiplications
Addition
Subtraction

This means that you work your way through a numerical problem by doing certain things first. For example:

$$10 \times 5 - 2 \quad \text{and} \quad 10 \times (5 - 2)$$

will give different answers. In the first case, we multiply the 10 and 5 to get 50, then subtract the 2 to get the answer 48. However, in the second case, we work out the bracket, to get 3, and then multiply by 10 to get the answer 30. With practice and confidence, you can take several steps at once, but if in doubt, just work out one thing at a time, writing down the intermediate answers, until you get the answer.

2.2 | ALGEBRA

Algebra uses letters to represent amounts or quantities which could be money, weight, people or whatever. Algebra allows us to develop formulae, or general statements about relationships between things, for example, if an agent gets a 10 per cent cut from the earnings of a singer, then you could express this as:

Agent's cut = 10 per cent of singer's fee

If we use single letters, say C for cut and F for fee and remember that 10 per cent is a tenth, we have an algebraic expression:

$$C = 0.1F$$

Obviously such expressions can get rather more complicated than this, but they are much shorter, and easier to use, than long sentences telling you what to do. You can apply the BEDMAS mnemonic in algebra too, but remember that you can only add things that are the same. For example:

$$4a + 6a + 2a \times 3b$$

Doing the multiplication gives $6ab$, and the addition gives $10a$, so the result is:

$$10a + 6ab$$

 Note that we don't usually bother with the multiplication sign between a number and a letter when writing down an algebraic expression.

2.3 | POWERS

When we have to multiply the same number by itself many times, we could end up with very long expressions. In this case we use **powers**. A power just tells us how many times the number is multiplied together. For example:

$$2 \times 2 = 2^2 = 4 \qquad \text{where 2 is the power}$$
$$2 \times 2 \times 2 = 2^3 = 8 \qquad \text{where 3 is the power}$$
$$a \times a \times a \times a = a^4 \qquad \text{where 4 is the power}$$

When the power is a 2, say 3^2 or b^2 we can talk about 3 squared or b squared. This is also known as **exponentiation** (the E in BEDMAS). What about negative powers? All this means is that we have one divided by the number raised to the power; so

$$2^{-2} = 1/2^2 = ¼$$
$$6 \times a^{-5} = 6/a^5$$

If we multiply a number raised to one power by the same number raised to another power, then we can just add the powers together, for example:

$$2^3 \times 2^4 = 2^{3+4} = 2^7 = 128$$
$$a^6 \times a^3 = a^{6+3} = a^9$$
$$a^4 \times a^{-3} = a^{4-3} = a^1 = a$$

So any number raised to the power of one is itself. Another special case is:

$$a^n \times a^{-n} = a^{n-n} = a^0 = 1$$

In other words, any number raised to the power of zero is equal to one.

We sometimes use fractional powers. These allow us to denote things like square roots or cube roots. For example:

$$\sqrt{4} = 4^{1/2} = 2$$
$$\sqrt[3]{27} = 27^{1/3} = 3$$

An obvious result from this is:

$$a^{1/2} \times a^{1/2} = a^{1/2+1/2} = a^1 = a$$

2.4 │ GRAPHS

A graph is useful since it gives us a picture to look at. This is often easier to understand than some complex equation. Just to remind you, a graph is plotted on a pair of axes, often labelled x and y, although any appropriate labels can be used. Where the two axes cross is called the **origin** and represents the point where both x and y are zero. The axes are at right angles to each other, as shown below in figure 2.1.

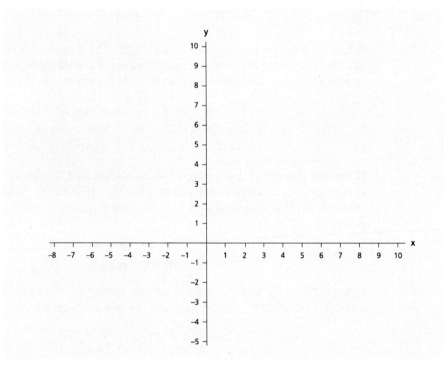

FIGURE 2.1   A blank graph

We spend most of our time working in the part of the graph where both x and y are positive, since, for many situations, negative values have no meaning, e.g. sales. Each point on the graph can be uniquely identified by two values, the x and the y. These are called **coordinates**. You are probably familiar with a similar idea for identifying cells in a spreadsheet.

Figure 2.2 shows a series of points and their coordinates on a graph.

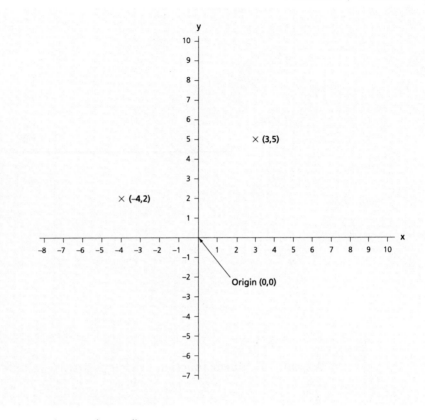

FIGURE 2.2   Points and coordinates

2.5 | FUNCTIONS

There are some standard functions, or equations, which will help in understanding many different situations and make calculations easier. We will look at a few of these here in terms of their structure, and their graphs.

(a) A **constant**: the simplest possible function, where something stays the same all of the time. The structure of the equation of a constant is:

$$y = k$$

A graph of a constant looks like the one in figure 2.3.

(b) A **linear function**: or more simply, a straight line. These are surprisingly powerful functions and find uses in accounting, statistics, economics and marketing, as well as the physical sciences. The structure of a linear function is:

$$y = a + bx$$

❗ You might have seen it written as $y = mx + c$ too.

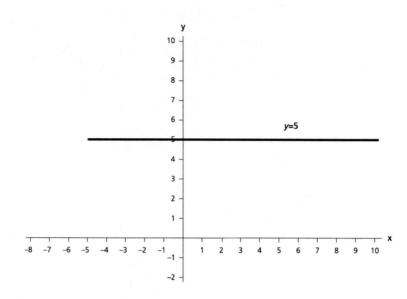

FIGURE 2.3   A constant

A graph of a linear function, with a positive b-value would look like figure 2.4:

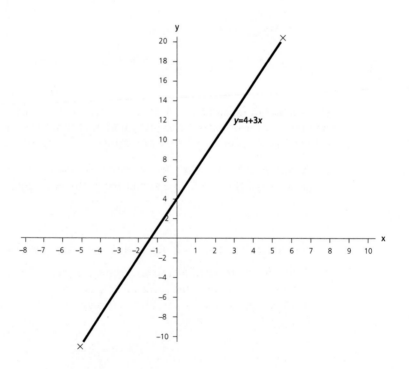

FIGURE 2.4   Linear function (*b* positive)

A graph with the *b*-value negative would look like figure 2.5:

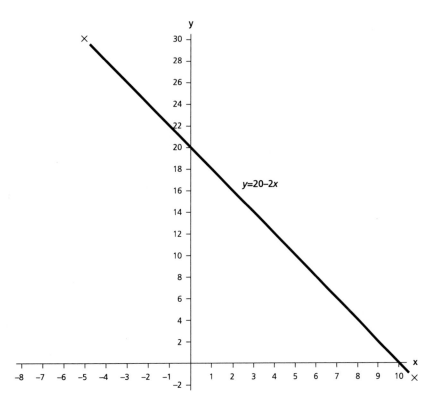

FIGURE 2.5   Linear function (*b* negative)

In order to draw a graph of a particular linear function we need to know the values for both a and b. The a-value is known as the **intercept** and is the place where the function cuts the vertical, or y axis. The b-value is known as the **slope**. Since the function is linear, we only need to know the coordinates of two points and then we can just join them together with a ruler.

For example, if we have the function $y = 10 + 3x$ we know the intercept is 10 so one pair of coordinates is (0,10). By putting an x-value into the equation we can find another pair of coordinates. Say we put in $x = 3$, then we get $y = 10 + 3 \times 3 = 19$, so the coordinates are (3,19); plotting these points on a graph and joining them up, gives figure 2.6.

As a second example we will use $y = 20 - 2x$. The intercept is 20 and the slope is –2. The first pair of coordinates are (0,20). Using $x = 5$, we get $y = 10$, so the second set of coordinates are (5,10). Joining up the points give the graph in figure 2.7.

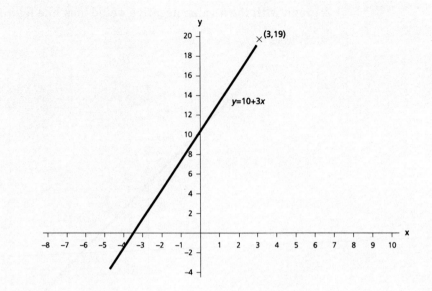

FIGURE 2.6   Graph of $y = 10 + 3x$

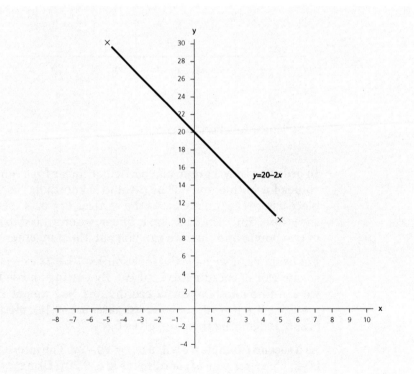

FIGURE 2.7   Graph of $y = 20 - 2x$

(c) **Quadratic function**: a quadratic function is one with a single bend in it. Again they turn out to be very useful in understanding a wide variety of situations, especially in economics. The structure of a quadratic function is:

$$y = ax^2 + bx + c$$

The shape of the graph of a quadratic function will depend upon the sign of the a-value in the equation. Where the a-value is positive, we get a graph like figure 2.8, but where it is negative, we get a graph like figure 2.9.

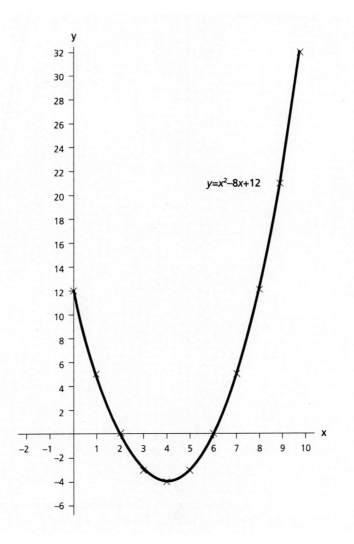

FIGURE 2.8   A quadratic with a positive a-value

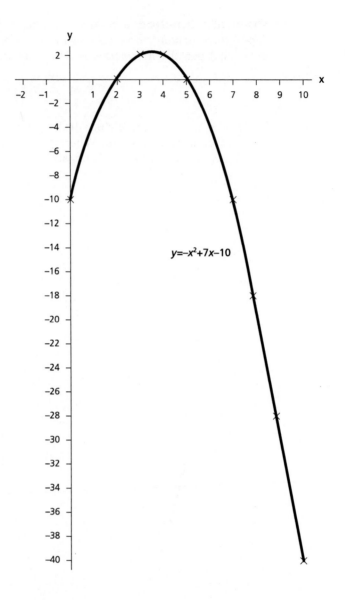

$$y=-x^2+7x-10$$

FIGURE 2.9   A quadratic with a negative *a*-value

In order to draw a graph of a quadratic, we need a fairly large number of points. For a very simple function, we could work these out by hand, but more complex functions can be calculated using a spreadsheet. As an example consider the function $y = -2x^2 + 5x + 120$.

To work out a series of coordinates we will use a spreadsheet and put each part of the function into separate cells as in figure 2.10. We can then add the various bits together to get the *y*-values of the coordinates.

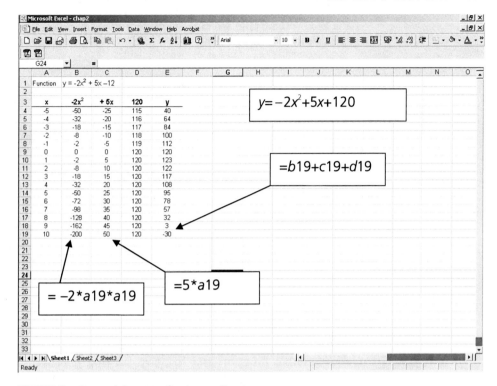

FIGURE 2.10   Spreadsheet to find coordinates

If you are not very familiar with Microsoft Excel, see the final section of the book where you will find a primer on that package. The same material is also available on the companion web site.

Plotting the various points and joining them together gives us figure 2.11.

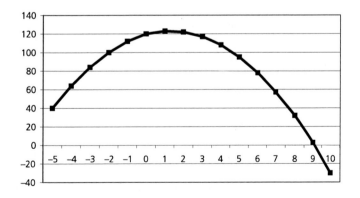

FIGURE 2.11   Graph of $y = -2x^2 + 5x + 120$

As a second example we can graph $y = x^2 - 5x + 10$. We get the following graph shown in figure 2.12:

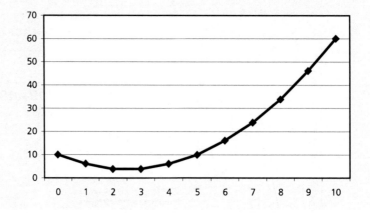

FIGURE 2.12   Graph of $y = x^2 - 5x + 10$

(d) Other functions: if you go on increasing the power of x you usually get functions with more and more bends, or turning points. If we have an x^3, it is called a cubic function and has two turning points. Cubic functions are sometimes used to represent total cost functions in economics. Where we have an x^4, it is called a quartic function, and has three turning points.

Finally there are two related functions which sometimes come up in financial calculations; these are the natural log and the exponential functions. At this stage we will just show their structure and illustrate their graphs. An exponential function uses a special number labelled e (actually equal to 2.71...), and the structure of the function is either $y = e^x$ or $y = e^{-x}$.

The graphs look like this one in figure 2.13:

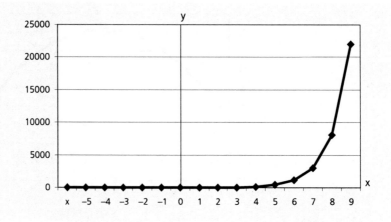

FIGURE 2.13   Exponential functions (a) $y = e^x$

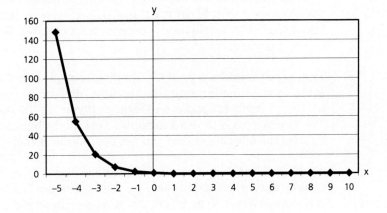

FIGURE 2.13   Exponential functions (b) $y = e^{-x}$

A natural log function is $y = \log_e(x)$ and the graph looks like the one shown in figure 2.14:

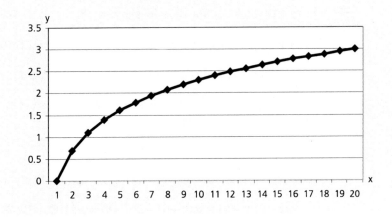

FIGURE 2.14   A log function

2.6 | ROOTS OF A QUADRATIC

You probably remember doing this at school, but perhaps can't quite remember how to do it. The good news is that you are very rarely expected to remember formulae in higher education, but it is amazing how many people can remember this one.

What is a root? A **root** is a place where a quadratic function crosses the x-axis. Earlier we said that a quadratic had one bend, or turning point, so if you think about it, not all quadratic functions will have roots! Some will drop in y-value as x increases, but never reach zero, and therefore not cross the x-axis. Others will increase in y-value as x increases, but always remain negative, so again, they do not cross the x-axis. Many functions, of course, will cross the x-axis in two places and therefore have two roots. Can you think of a situation where a quadratic has one root (twice)?

There are basically two ways of finding roots, if they exist. The first consists of splitting the function into two brackets, multiplied together. This often works for fairly simple functions and is quick if you see the roots (i.e. guess correctly). The alternative is to use a formula. This might be slower for simple functions, but has the advantage that you get the answer if there is one.

Looking at brackets first, we will take $y = x^2 - 8x + 12$ as our example. (It is usually only worth trying this method if the coefficient of x^2 is one, as in this case.) The roots will be where $y = 0$, so we can write:

$$x^2 - 8x + 12 = 0$$

Our answer is going to look like this:

$$(x + a)(x + b) = 0$$

where a times b equals 12, and $a + b = -8$.

Well, the pairs of numbers which give 12 when multiplied together are:

12 and 1, 6 and 2, 4 and 3, –12 and –1, –2 and –6, and –3 and –4

But the only pair which add to –8 are –2 and –6, so these must be the values of a and b. Therefore the result is:

$$(x - 2)(x - 6) = 0$$

Just to show that our answer works, we can multiply out the brackets:

$$(x - 2)(x - 6) = x(x - 6) - 2(x - 6) = x^2 - 6x - 2x + 12 = x^2 - 8x + 12$$

For that to be true, either the first bracket, or the second one, must be equal to zero. So:

If $(x - 2) = 0$, then $x = 2$ is one root.
If $(x - 6) = 0$, then $x = 6$ is the other root.

We can check that this is true by substituting these answers into the original equation:

If $x = 2$, we have $(2)^2 - 8(2) + 12 = 4 - 16 + 12 = 0$
If $x = 6$, we have $(6)^2 - 8(6) + 12 = 36 - 48 + 12 = 0$

We advise you to draw the function's graph, as in figure 2.15 to reinforce this concept.

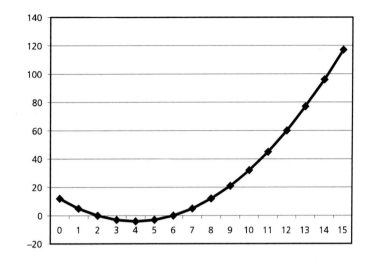

FIGURE 2.15   Graph of $y = x^2 - 8x + 12$

Taking a second example, if we are looking for the roots of:

$$y = x^2 - x - 12$$

we put the equation equal to zero:

$$x^2 - x - 12 = 0$$

Pairs of numbers which multiply to give –20 are:

+1 and –20, +2 and –10, +4 and –5, +5 and –4, +10 and –2, +20 and –1

but the only pair that added together give an answer of –1 is +4 and –5

so $(x + 4)(x - 5) = 0$

checking, we have:

$$x(x - 5) + 4(x - 5) = x^2 - 5x + 4x - 20 = x^2 - x - 20$$

So, the roots are at:

$x + 4 = 0$, giving $x = -4$

and $x - 5 = 0$, giving $x = 5$

To answer our earlier question, a quadratic will have the same root (twice) if it just touches the x-axis. Take for example, $y = x^2 - 10x + 25$. Here we get two brackets:

$$(x - 5)(x - 5) = 0$$

If $(x - 5) = 0$, then $x = 5$, and the same for the other bracket. This would give the graph in figure 2.16.

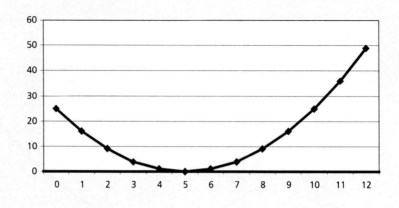

FIGURE 2.16   Graph of $y = x^2 - 10x + 25$

The alternative way of finding roots is to use the formula. This uses the letters from the general equation we gave earlier:

$$y = ax^2 + bx + c$$

and the roots are at:

$$\frac{-b - \sqrt{(b^2 - 4ac)}}{2a}$$

If roots exist, this will find them. We will work through one example of using this formula, but suggest that you try all of the ones in the exercises. We will use

$$y = -2x^2 - 4x + 30$$

Here $a = -2$, $b = -4$ and $c = 30$, so we get:

$$\frac{+4 - \sqrt{16 + 240}}{-4}$$

$$\frac{+4 - \sqrt{256}}{-4} = \frac{+4 + 16}{-4} \text{ or } \frac{+4 - 16}{-4} = \frac{+20}{-4} \text{ and } \frac{-12}{-4}$$

so the roots are –5 and 3. Try constructing the graph to show that this is the case. Obviously not all roots turn out to be whole numbers.

2.7   SIMULTANEOUS EQUATIONS

Other subject areas may require you to solve **simultaneous equations**. In most cases, the numbers are fairly straightforward. We usually have two equations, both of which are straight lines. The solution to them as simultaneous equations is where they cross on a graph, but it is usually simpler to use algebra rather than draw a graph to find the solution.

Given two simultaneous equations we make the coefficient of either the x or the y equal in both equations (by multiplying one or both equations by convenient numbers) and then subtract or add the equations. From here we can find either the x or the y value. Having got one answer, the other can be found by substitution. An example will help!

$$2x + 5y = 26 - \text{equation 1}$$
$$x + 10y = 43 - \text{equation 2}$$

If we multiply the first equation by 2, the coefficients of y will become equal:

$$4x + 10y = 52$$
$$x + 10y = 43$$

now we subtract the second from the first, which gives:

$$3x = 9$$

Divide by 3, giving:

$$x = 3$$

Now we know this, we can substitute $x = 3$ into the first equation and get:

$$6 + 5y = 26$$
$$5y = 20$$
$$y = 4$$

So the two equations are both true at $x = 3$ and $y = 4$. If we were to draw their graphs, they would cross at the point (3,4); see figure 2.17.

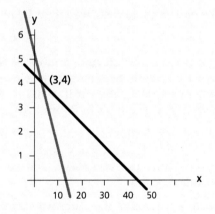

FIGURE 2.17   Simultaneous equations

Looking at a second example:

$$3x + 4y = 19 \qquad \text{equation 1}$$
$$5x + 3y = 6 \qquad \text{equation 2}$$

If we multiply equation 1 by 3 and equation 2 by 4, we get:

$$9x + 12y = 57 \qquad \text{equation 3}$$
$$20x + 12y = 24 \qquad \text{equation 4}$$

Now subtracting equation 4 from equation 3, we get:

$$-11x = 33$$
$$x = -3$$

Substituting into equation 1, we get:

$$3(-3) + 4y = -9 + 4y = 19$$
$$4y = 28$$
$$y = +7$$

So the simultaneous equations are both true at $x = -3$ and $y = +7$.

2.8 | INEQUALITIES

So far we have dealt with equations; that is, situations where one side of the relationship is exactly equal to the other side. This is not the only type of relationship. Sometimes we know, or suppose, that one side is less than the other side. These types of relationship can be very important in describing and analyzing situations, and we illustrate this in the next chapter. Here we will look at two particular forms of inequality and the way that they might be illustrated.

We will only look at inequalities with one or two variables.

For the one variable case, we might have:

$$X < 10$$

This means that for all time and in all situations, the variable X is below the value of 10. It could be a long way below, for example, -200, or it could be just below, for example, 9.99, but it remains below 10. This inequality can be shown by an area on a graph, as in figure 2.18.

There is a very similar case which is written as $X \leq 10$ which means that X is less than or equal to 10. Fairly obviously, in this case X can take the value of 10, but no more. Whilst the two cases look similar, the difference between them is important.

The inequality can be the other way around, as in:

$$X \geq 10$$

This is illustrated in figure 2.19, but note that the line is included in the acceptable values this time; previously it was just the boundary.

Where we have two variables we can also illustrate an area represented by the inequality, for example, if we have:

$$2X + 5Y \leq 200$$

then the first step is to consider the equation part, i.e. $2X + 5Y = 200$. From this we can find the value of X if Y is zero (100,0) and Y when X is zero.

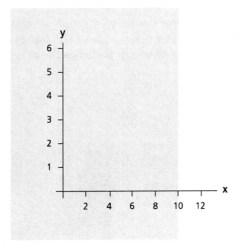

FIGURE 2.18   Area for $X < 10$

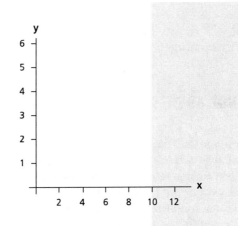

FIGURE 2.19   Area for $X \geq 10$

$$2X + 0 = 200, \text{ so } X = 100$$

and the value of Y if X is zero:

$$0 + 5Y = 200, \text{ so } Y = 40$$

We now have two sets of coordinates for our graph (100,0) and (0,40).

If we plot these points on a graph we can join them up with a straight line. Now we can consider which side of the line is consistent with the inequality – here it will be that side which includes the origin. This is shown in figure 2.20.

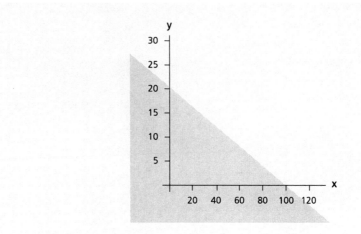

FIGURE 2.20   Area for $2X + 5Y \leq 200$

We will return to inequalities in the next chapter when we consider linear programming.

2.9 | FREQUENCY COUNTS

You wouldn't normally find this section in such a chapter as this, but we feel that it is a basic part of using numbers successfully to understand problems. The situation which often arises is that the same numbers, or answers, come up many times, and we can count up how many times each one occurs. The number of times something occurs is called its **frequency**. This is often the case if you have conducted a survey and want to know how many people gave each of the answers to a question. There is nothing especially clever or difficult about this, but it does make the data easier to understand.

Let's say we have a survey where people were asked if they had ever taken an illegal substance. Further, we will give the label 1 to those who said 'Yes', and the label 2 to those who said 'No'. Our results might look like Table 2.1:

1	2	1	2	1
2	2	1	2	2
2	2	2	1	2
1	2	1	2	1
2	1	1	1	2
2	2	1	2	2

TABLE 2.1 Survey of people who have taken an illegal substance

This may be the raw data, but it fails to communicate the results of the survey. However, if we count up the number of 1s and 2s, we get the following information shown in table 2.2:

Answer	Code	Frequency
Yes	1	12
No	2	18

TABLE 2.2 Results when raw data is processed

This gives us a much better idea about what was said. You can do this by hand, but it is also possible to make a machine do it for you. For a small survey with just a few questions, Excel will find the frequency counts for you. If you carry out a more extensive survey and need to do more complex analysis, then it is worth putting the data into a specialist package such as SPSS (the Statistical Package for the Social Sciences).

2.10 PERCENTAGE CHANGES

When we want to make comparisons between two or more sets of numbers we often run into difficulties when there are big differences in size between the sets. If two sets both start off at 20 and the second value on one set is 25 and on the other is 50, it is easy to say which has increased the most! However, if one set starts at 20 and then goes to 25 whilst the other set starts at 5000 and then goes to 5100, it is less obvious which has the largest increase. One way to overcome this is to use **percentage changes** in the numbers.

Year	Amount
1	20
2	25
3	30
4	33
5	40

TABLE 2.3 Raw database using percentage change formula

To do this we use the initial value of the set as our starting point. We then need to find the change from the first number to the second, and finally make this a percentage of the starting point. Look at the example in table 2.3.

Taking Year 1 and Year 2, we have a change of (25 – 20) = 5 from a starting point of 20. To get a percentage, we put the change over the starting point and multiply the result by 100:

$$\text{Percentage change} = \frac{(25 - 20)}{20} \times 100 = \frac{5}{20} \times 100 = 25 \text{ per cent}$$

We then move on to Year 2 and Year 3:

$$\text{Percentage change} = \frac{(30 - 25)}{25} \times 100 = \frac{5}{25} \times 100 = 20 \text{ per cent}$$

We could continue by hand, but for most purposes you would use a spreadsheet.

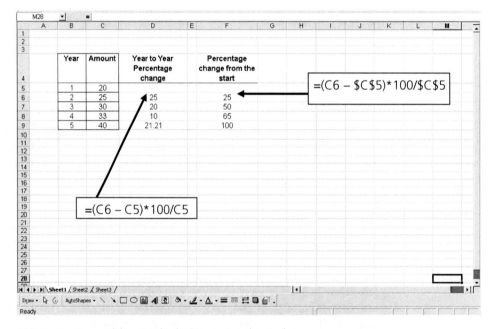

FIGURE 2.21   Spreadsheet calculating percentage changes

As you can see on the spreadsheet, you can also find the percentage change from a particular starting point, in our case, Year 1. Using year to year percentage changes is particularly useful when the sets of data are measured in different units, for example, different currencies. Percentage changes form the basis of index number calculations (see Chapter 7).

2.11 | CONCLUSIONS

This chapter has reviewed, briefly, the basic arithmetic and algebra that you are likely to need on a first-year course. Many courses have now made Quants a single-semester module and often expect you to be able to do all of these things before you start. We have not been able to go into depth in these topics, but many people just need a reminder about a few things. If you need to work through more slowly and from a slightly more basic level, then we suggest that you read Jon Curwin and Roger Slater, *Improve Your Maths: A Refresher Course* (Thomson Learning, 2000).

2.12 EXERCISES

1 Work out $\qquad$ $12 \times 4 + 6 - (3 \times 5)$

2 Evaluate $\qquad$ $(5 + 3) \times (4 - 1)$

3 Evaluate $\qquad$ $10 \times 4 - 3 + (2 \times 6) - (12 + 7)$

4 Simplify this expression: $\qquad$ $3a + 5a - (2a \times 5b) + 2a\,(2 + 5)$

5 Simplify $\qquad$ $3\,(2a + 5a) - 2\,(2a - a)$

6 Simplify $\qquad$ $4xy - 5x + 3y - 2x(2y - 5 + y)$

7 Simplify $\qquad$ $2^3 \times 2^4 \times 2^{-5}$

8 Simplify $\qquad$ $a^2 \times a^4 \times a^3 \times a^{-9}$

9 Simplify $\qquad$ $a^3 + a^2$

10 Construct a graph of $\quad y = 20$

11 Construct a graph of $\quad y = 25 + 3x$

12 Construct a graph of $\quad y = 50 - 10x$, and $y = 30 - 5x$ on the same axes.

13 Graph the function $\quad y = x^2 - 5x + 4$

14 Construct a graph of $\quad y = -2x^2 + 15x - 10$

15 Find the roots of $\quad y = x^2 - 12x + 20$

16 Find the roots of $\quad y = x^2 + x - 30$

17 Find the roots of $\quad y = x^2 - 4x - 21$

18 Find the roots of $\quad y = 2x^2 + 16x + 30$

19 Find the roots of $\quad y = 3x^2 + 5x - 10$

20 Find the roots of $\quad y = -2x^2 - 4x + 20$

21 Solve the following pair of simultaneous equations: $\quad 2x + 3y = 29$
$\qquad\qquad\qquad\qquad\qquad\qquad\qquad\qquad\qquad\qquad 4x + 5y = 55$

22 Solve the following pair of simultaneous equations: $\quad x - 4y = 1$
$\qquad\qquad\qquad\qquad\qquad\qquad\qquad\qquad\qquad\qquad\qquad 2x + 2y = 12$

23 Solve the following simultaneous equations: $\qquad\quad 10x + 2y = 34$
$\qquad\qquad\qquad\qquad\qquad\qquad\qquad\qquad\qquad\qquad\qquad 3x + 5y = 30$

24 Solve the following simultaneous equations: $\qquad\quad 4x + 6y = 8$
$\qquad\qquad\qquad\qquad\qquad\qquad\qquad\qquad\qquad\qquad\qquad 3x + 4y = 5$

25 Solve the following simultaneous equations: $\qquad\quad 2x - 5y = 31$
$\qquad\qquad\qquad\qquad\qquad\qquad\qquad\qquad\qquad\qquad\qquad 4x - 3y = 13$

26 Illustrate the following inequalities on graphs:

a $4x + 5y \geq 200$

b $x + 2y \geq 40$

c $2x + 3y \leq 60$

d $x \geq 20$

e $y \leq 25$

27 Create a frequency table from the following raw data.

1	3	3	2	2	2	1	1
2	1	2	1	1	2	1	2
3	3	1	2	3	3	2	2
3	1	2	3	1	3	2	1
2	1	1	2	2	3	3	3
2	2	3	2	1	1	1	3

28 People entering a shopping centre are counted at 10.00, 11.00, 12.00, 13.00, 14.00, 15.00 and 16.00 each day. Each observation is from a still off the CCTV system over the entrance. Results are as follows:

Day/Time	10	11	12	13	14	15	16
Mon	20	25	22	28	27	22	18
Tues	25	24	27	30	25	20	21
Wed	23	26	26	29	25	24	19
Thur	24	28	29	33	27	22	20
Fri	26	30	32	36	32	28	25
Sat	32	40	43	45	42	40	32
Sun		10	40	41	30	24	
Mon	21	22	24	29	26	22	20
Tues	25	24	28	32	27	23	19

Continued

Day/Time	10	11	12	13	14	15	16
Wed	24	27	27	31	26	21	20
Thur	26	29	31	33	28	24	21
Fri	29	35	35	38	33	30	24
Sat	40	45	44	45	45	39	36
Sun		9	45	45	29	27	

The centre is only open from 11.00 to 16.00 on Sundays.

Create a frequency table from this data and then create a grouped frequency table which you could use in a presentation.

29 For the following table of data

a   Find the percentage change from Year 1.

b   Find the year to year percentage changes in the data.

Year	Amount sold
1	40
2	42
3	45
4	50

30 Go to the government web site at **http://www.statistics.gov.uk** and download data on Gross National Income: Current prices: NSA (ABMX). This will give you a file you can open with Excel. Use this data to find the year to year percentage changes and then construct a graph of these figures.

Note that year to year percentage changes might also be called growth.

2.13 ANNOTATED ANSWERS

1   Here we use the BEDMAS rule, so do the two multiplications first.

This gives: 48 + 6 – 15, then we can do the addition and subtraction to get the answer of **39**

2   BEDMAS again, work out the brackets first to get 8 × 3 , and then do the multiplication to get the answer **24**

3   This one is a bit longer, but the same idea. Work out the brackets and the multiplication to give 40 – 3 + 12 – 19 , and then do the addition and subtraction to get the answer of **30**

4 This is BEDMAS with algebra too. Work out the brackets and multiplications to give: $8a - 10ab + 14a$, then add like terms together giving

$$22a - 10ab$$

5 Work out the brackets to give: $3 \times 7a - 2 \times a$, then do the multiplications giving $21a - 2a$, finally do the subtraction to get the answer **$19a$**

6 First simplify the bracket and then multiply it by the $2x$ to give: $4xy - 5x + 3y - 4xy + 10x - 2xy$, and then add together like terms for the answer:

$$5x + 3y - 2xy$$

7 We just add the powers together giving 2^{3+4-5} so the answer is

$$2^2$$

8 Since the terms are multiplied, we add the powers $a^{2+4+3-9}$, but when we do this we get a^0. But anything to the power zero is equal to **1**

9 This is a trick question, the terms are added together so the answer is

$$a^3 + a^2 \qquad \text{(it can't be simplified)}$$

10 Here we are graphing a constant, so y will be equal to 20, no matter what value of x we use. This gives us a horizontal line, as in the figure below:

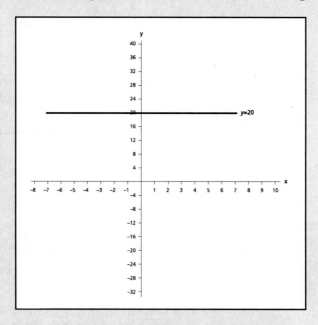

11 This is a simple linear function where y increases by 3 each time x increases by one. The intercept (where $x = 0$) is at $y = 25$.

This gives:

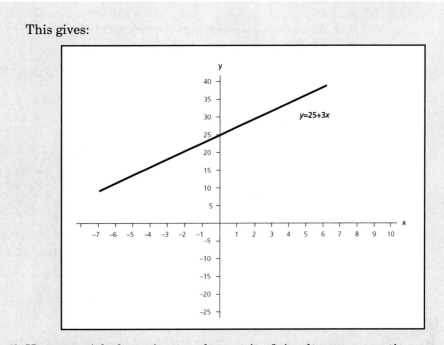

12 Here we might be trying to solve a pair of simultaneous equations or a
linear programming problem. These two functions give us the graph
below:

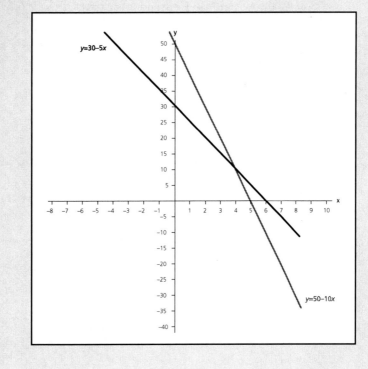

13 We can use a spreadsheet to construct this graph:

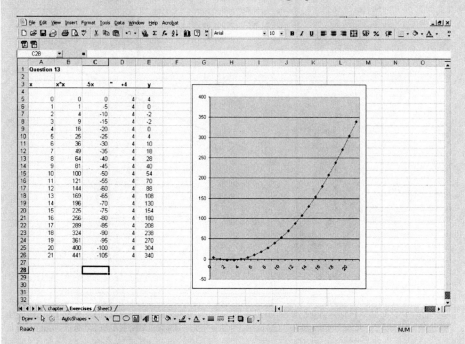

x	x*x	-5x	+4	y
0	0	0	4	4
1	1	-5	4	0
2	4	-10	4	-2
3	9	-15	4	-2
4	16	-20	4	0
5	25	-25	4	4
6	36	-30	4	10
7	49	-35	4	18
8	64	-40	4	28
9	81	-45	4	40
10	100	-50	4	54
11	121	-55	4	70
12	144	-60	4	88
13	169	-65	4	108
14	196	-70	4	130
15	225	-75	4	154
16	256	-80	4	180
17	289	-85	4	208
18	324	-90	4	238
19	361	-95	4	270
20	400	-100	4	304
21	441	-105	4	340

14 Again we can use a spreadsheet to get this graph:

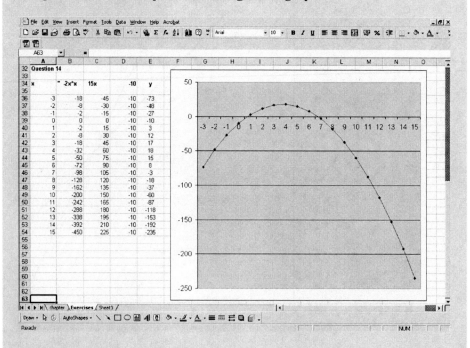

x	-2x*x	15x	-10	y
-3	-18	-45	-10	-73
-2	-8	-30	-10	-48
-1	-2	-15	-10	-27
0	0	0	-10	-10
1	-2	15	-10	3
2	-8	30	-10	12
3	-18	45	-10	17
4	-32	60	-10	18
5	-50	75	-10	15
6	-72	90	-10	8
7	-98	105	-10	-3
8	-128	120	-10	-18
9	-162	135	-10	-37
10	-200	150	-10	-60
11	-242	165	-10	-87
12	-288	180	-10	-118
13	-338	195	-10	-153
14	-392	210	-10	-192
15	-450	225	-10	-235

15 We could draw a graph, but it is quicker to break the expression down into two brackets. We want two numbers that when multiplied give us +20, and when added give us –12. The pair that fit these criteria are –10 and –2, so we get $(x - 10)(x - 2) = 0$, so the roots are at $(x - 10) = 0$ and $(x - 2) = 0$. This gives the two answers:

$$x = 2 \text{ and } 10$$

16 Here the two numbers need to add up to +1 and give –30 when multiplied. The pair that do this are –5 and +6. So the brackets are $(x - 5)(x + 6) = 0$, so the roots are at:

$$x = 5 \text{ and } –6$$

17 Here we need a sum of –4 and a product of –21 for our two numbers. This gives the pair of brackets $(x - 7)(x + 3) = 0$, and so the roots are at:

$$x = 7 \text{ and } –3$$

18 We could use brackets here, but the $2x^2$ may make it tricky so we will use the formula. Looking at the quadratic, we can identify that $a = +2$, $b + 16$, and $c = +30$. These can now be substituted into the formula to give:

$$\frac{-16 - \sqrt{256 - 240}}{4} = \frac{-16 - 4}{4} = –5 \text{ and } –3$$

19 Again the formula is the simplest approach to this question.

We have $a = +3$, $b = +5$ and $c = -10$. Substituting into the formula gives:

$$\frac{-5 - \sqrt{25 - 120}}{6} = \frac{-5 - 12.04}{6} = –2.84 \text{ and } 1.173$$

20 Here we can identify $a = –2$, $b = –4$ and $c = +20$. Substituting gives:

$$\frac{4 - \sqrt{16 + 160}}{-4} = \frac{4 - 13.2665}{-4} = –4.317 \text{ and } 2.317$$

21 We need to get a pair of coefficients to be the same so that we can eliminate one of the variables. One way is to multiply everything in the first equation by 2 to give us:

$$4x + 6y = 58$$
$$4x + 5y = 55$$

If we now subtract the second equation from the first, we get:

$$y = 3$$

Now substitute this value into one of the equations, say the second one, to get:

$$4x + 15 = 55$$

So

$$4x = 40 \text{ and } x = 10$$

The answer is:

$$\boldsymbol{x = 10, y = 3}$$

22 For this question we can make the coefficients of x both equal to 2 by multiplying the first equation by 2:

$$
\begin{aligned}
2x - 8y &= 2 \\
\underline{2x + 2y} &= \underline{12} \\
-10y &= -10 \\
y &= 1
\end{aligned}
$$

Substituting in the first equation gives:

$$x - 4 = 1, \text{ so } x = 5$$

The answer is:

$$\boldsymbol{x = 5, y = 1}$$

23 To equate the coefficients of y we can multiply the first equation by 5 and the second one by 2 to give:

$$
\begin{aligned}
50x + 10y &= 170 \\
\underline{6x + 10y} &= \underline{60} \\
44x &= 110 \\
x &= 2.5
\end{aligned}
$$

Substituting into the first equation gives:

$$25 + 2y = 34, \text{ so } 2y = 9, \text{ and } y = 4.5$$

The answer is:

$$\boldsymbol{x = 2.5, y = 4.5}$$

24 To equate the coefficients of x we can multiply the first equation by 3 and the second by 4

$$
\begin{aligned}
12x + 18y &= 24 \\
\underline{12x + 16y} &= \underline{20} \\
2y &= 4 \\
y &= 2
\end{aligned}
$$

substituting in the first equation gives:

$$4x + 12 = 8, \text{ so } 4x = -4, \text{ and } x = -1$$

The answer is:

$$\boldsymbol{x = -1, y = 2}$$

25 To equate the coefficients of x we can multiply the first equation by 2:

$$
\begin{aligned}
4x - 10y &= 62 \\
4x - 3y &= 13
\end{aligned}
$$

$$-7y = 49$$
$$y = -7$$

substituting in the second equation gives:

$$4x - 3(-7) = 4x + 21 = 13, \text{ so } 4x = -8, \text{ and } x = -2$$

The answer is:

$$x = -2, y = -7$$

26

a

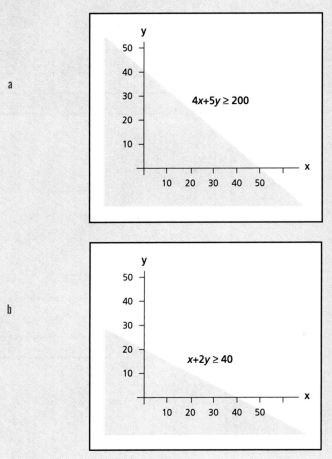

b

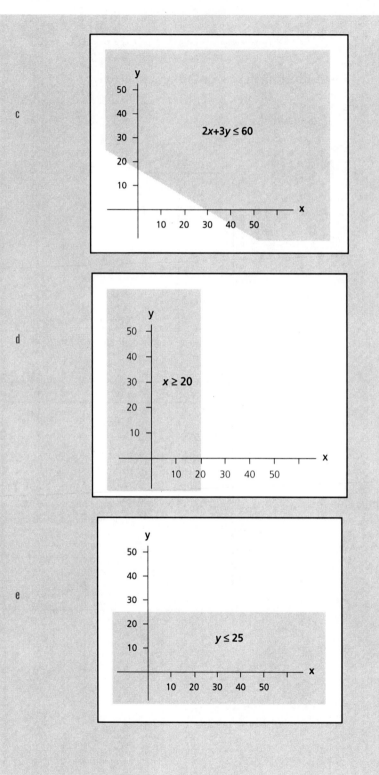

27 The first thing to notice here is that the only numbers that appear in the table are 1, 2 and 3. We can then just count up the number of times each one occurs, to give the answer. The frequency table is:

Code	Frequency
1	16
2	18
3	14

We could have used a spreadsheet to do this, but it takes longer to type the data in than it does to count up the numbers. If you already have the data in electronic form and are going to go on and do more analysis, then the spreadsheet will get to the answer more quickly.

28 This question is more complicated since the range of numbers in the table is much wider, from 9 to 45. The first step is to create a tally chart of the numbers. This is just a way of counting up:

X		X		X		X	
9	/	19	//	29	//////	39	/
10	/	20	/////	30	////	40	////
11		21	////	31	//	41	/
12		22	/////	32	/////	42	/
13		23	//	33	///	43	/
14		24	/////////	34		44	/
15		25	//////	35	//	45	//////
16		26	//////	36	//		
17		27	////////	37			
18	/	28	/////	38	/		

Counting up the ticks gives the frequency table:

X	f	X	f	X	f	X	f
9	1	19	2	29	6	39	1
10	1	20	5	30	4	40	4
11		21	4	31	2	41	1
12		22	5	32	5	42	1
13		23	2	33	3	43	1
14		24	9	34		44	1

Continued

X	f	X	f	X	f	X	f
15		25	6	35	2	45	5
16		26	6	36	2		
17		27	7	37			
18	1	28	5	38	1		

To make this table suitable to use in a presentation we need to group the values together. There is no correct answer, but we usually try to make the groups of similar widths unless there are a few extreme values. We chose the following group sizes:

Number of people entering	Frequency (f)
Under 20	5
20 but under 25	25
25 but under 30	30
30 but under 35	14
35 but under 40	6
40 or more	14
Total	94

29 For the percentage changes from Year 1 we subtract the Year 1 figure (40) from each figure, multiply by 100 and then divide by the Year 1 figure – so for Year 2 we have:

$$(42 - 40) \times 100/40 = 200/40 = \ 5 \text{ per cent}$$

$$\text{Year 3} \quad (45 - 40) \times 100/40 = 500/40 = 12.5 \text{ per cent}$$

$$\text{Year 4} \quad (50 - 40) \times 100/40 = 1000/40 = 25 \text{ per cent}$$

For year to year changes we subtract last year's figure from this year's, multiply the result by 100 and then divide by last year's figure – so for Year 2 we have:

$$(42 - 40) \times 100/40 = 200/40 = \ 5 \text{ per cent}$$

$$\text{Year 3} \quad (45 - 42) \times 100/42 = 300/42 = \ 7.14 \text{ per cent}$$

$$\text{Year 4} \quad (50 - 45) \times 100/45 = 500/45 = 11.11 \text{ per cent}$$

30 Answers to this question will vary depending on how much data you download and exactly when you do so. Our answer is based on a download in July 2003 and is shown in the following screen shot.

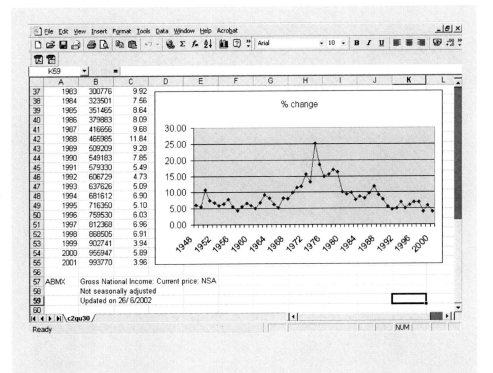

See the companion web site for further questions and annotated answers. There you will also find a PowerPoint presentation which you can use to help understand this area, or later, for revision. The site also contains links to other sites of interest.

Chapter 3

USING GRAPHS

The basics of using algebra and drawing graphs may be useful for classroom exercises, but we need to look at ways in which they can be used to solve, or at least give an insight into, problems from a range of subject areas. This chapter looks at some of the most useful ways of using graphs.

OBJECTIVES

After reading this chapter and doing the exercises you should be able to:

- create graphs of time series data
- illustrate break-even analysis
- show a feasible area
- solve two variable LP maximization problems
- solve two variable LP minimization problems
- discuss the practical use of break even and LP

3.1 | TIME SERIES GRAPHS

One of the simplest, but very useful ways of using a graph is to show what has happened over time. Usually, the longer the period, the better, unless something has happened to completely change the data. (This sometimes happens with a change in definition of some published data.) All of the normal rules apply: the graph should have a title, the axes should be clearly labelled and the units of measurement should be clearly stated.

Such graphs can be drawn by hand, but in most cases you would use Excel. In many ways this makes life easier, but the need to take care of labelling is even more important. One reason is that time series data may be annual, quarterly, monthly, weekly, daily or even hourly. In Chapter 9 we will develop analytical methods for dealing with **time series** data; here we are attempting to get a more intuitive feel for the data.

Annual data: Where we have annual data, constructing the graph should give a relatively smooth picture. What we are looking for is an indication of broad underlying movements in the figures which might help to explain changes in the past and suggest what will happen in the future. A typical graph is shown in figure 3.1.

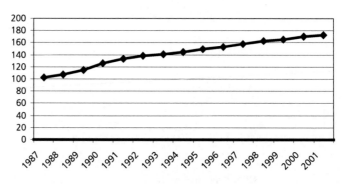

Retail prices – annual data

FIGURE 3.1   Annual time series data taken from **http://www.statistics.co.uk**

This is a long-running series which shows a general upward trend. We will discuss the retail prices index further in Chapter 7.

Quarterly data: Where we have quarterly data, the constructed graph will often have a much more jagged appearance. What we have here is a combination of both the overall movement in the data from year to year, and individual **seasonal effects**. A seasonal effect is something like selling more electricity in the winter months, or more Easter eggs at Easter. What we are looking for with such data is whether or not the quarterly movements are about the same in the same quarter each year. See figure 3.2 for quarterly time series data.

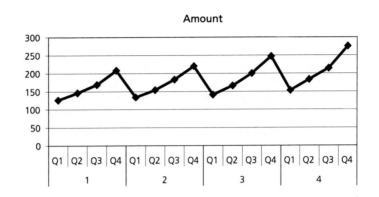

FIGURE 3.2   Quarterly time series data

Figure 3.2 shows stylized data, but illustrates that the data is generally increasing over time and has a pronounced seasonal pattern.

The other sorts of data could also be illustrated, but constructing the graphs from Excel allows you to see the patterns fairly quickly. However, if you work with a very large amount of data, the enforced small scale may hide some of the effects we are looking for, so it may be necessary to expand the chart box within the Excel spreadsheet.

Time series data can often be downloaded, especially from the National Statistics web site at **http://www.statistics.gov.uk**. This will often be your starting point if you are investigating changes over time. Obtaining data from companies can be slightly more difficult, but short runs of data are often contained in annual reports (also often on the companies' web sites). You might also try the site at **http://www.carol.co.uk**, where you can register (for free) to look at company reports.

3.2 | BREAK-EVEN ANALYSIS

A fundamental question in running any business or selling a product or service is 'Will it make money?'. Even where the objective is not necessarily to make a profit, in, for example, charitable activities, it is often important to cover costs. In other words, to **break even**. The tools we developed in Chapter 2 may allow us to answer the question; in fact, they allow us to go further, and suggest the level, or levels of output which will show a break-even result. Obviously, if we don't break even, we either make a profit or a loss!

Let's start by looking at a simple situation. Suppose we have a product which we can sell for 5. (This could be in pounds, euros, dollars, or any other currency.) If we sell 10 items, we receive 50 (i.e. 5×10); if we sell 100, we get 500 (i.e. 5×100). More generally, if we sell X units, we receive $5X$. This is all the money we get, so it is called **total revenue** (TR).

We can now look at costs. Suppose the items cost us 3 to make. (This will be in the same units or currency as the selling price.) Using the same logic we can say that our cost of production, if we make X units, are $3X$ (usually called **variable costs**). Businesses will also have a **fixed cost**, which they have to pay whether or not they have any production (things like rent, business rates, etc.). For our company, let's say that their fixed costs are equal to 200. Therefore the **total costs** of the business are:

$$TC = 200 + 3X$$

Break even is where $TR = TC$, so:

$$
\begin{aligned}
TR &= TC \\
5X &= 200 + 3X \\
2X &= 200 \\
X &= 100
\end{aligned}
$$

So the break-even level of production is 100 units. You might like to note that in simple examples such as this, the level is equal to the fixed cost divided by the **contribution**, where contribution is the difference between selling price and cost per unit. We can illustrate the situation using a graph, as in figure 3.3.

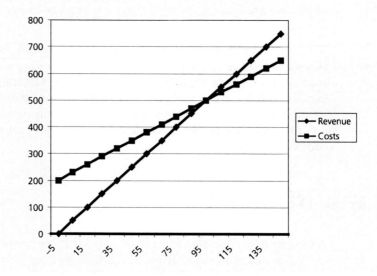

FIGURE 3.3   Break-even chart

Just before we move on, you could also note that we can define profit as $TR - TC$, so here we have:

$$\text{Profit } (\Pi) = 5X - (200 + 3X) = 2X - 200$$

And we could have drawn the graph of this profit function and noted where it is equal to zero. Since zero profit is break even, this will also solve the problem. This is illustrated in figure 3.4.

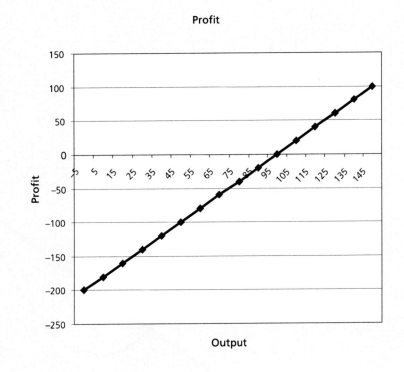

FIGURE 3.4   Linear profit function

As you can see, all three methods give the same answer. The profit approach may be favoured by economists, whilst the contribution approach is more popular with accountants.

Of course, not many cost functions are **linear** (straight lines). Many will be 'U-shaped', which we discovered in the last chapter could be represented by a **quadratic function**. This will also make the profit function quadratic. It seems unlikely that the total revenue function for a single product will be anything but linear. (This may not be an assumption that you can make when dealing with much more complex models.)

Suppose that we have a product which we can sell at 20 each. This gives a total revenue of $20X$.

Suppose also that total cost is represented by the equation

$$TC = X^2 - 100X + 3000$$

If we use a spreadsheet to draw these two functions, we will get the diagram in figure 3.5:

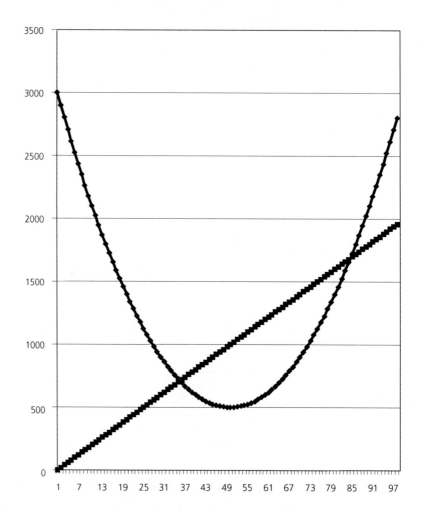

FIGURE 3.5   Total revenue and total cost functions

As you can see, the total revenue line cuts the total cost line twice, giving two break-even points. Below the first point we make a loss. Between the two points we make a profit, since revenue is above costs. Above the second point we make a loss. The company could choose to operate at any point between the two break-even points. It may, of course, have no control over the amount it sells, but this tells us that provided the sales are between the two points, the company will be profitable.

As before, it may be easier to construct a profit function:

$$\text{Profit } (\Pi) = TR - TC$$
$$= 20X - (X^2 - 100X + 3000)$$
$$= 20X - X^2 + 100X - 3000$$
$$= -X^2 + 120X - 3000$$

This is graphed in figure 3.6.

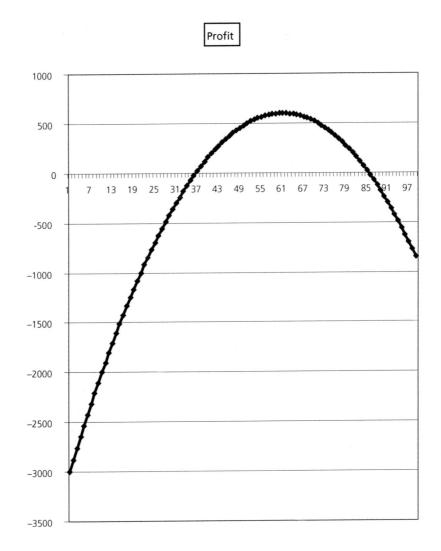

FIGURE 3.6   Profit function (non-linear)

Looking at figure 3.6, you should recognize that the break-even points are the **roots** of the profit function, so we could use the formula introduced in the previous chapter to find the values, instead of reading them directly from the graph.

If this is done, we get the values 35.5 and 84.5. Maximum profit looks to occur at about $x = 60$, according to figure 3.6.

❗ Note that it is beyond the scope of this book to determine algebraically the output level to give maximum profit, but we can work from the graphs and the spreadsheets to estimate the levels for maximum profit. If you wish to see the solution using calculus, then see Jon Curwin and Roger Slater, *Quantitative Methods for Business Decisions* (Thomson Learning, 2002).

Break-even analysis is relatively straightforward for a single product company, but it is also used in multi-product companies. In these cases, the fixed costs are a proportion of the firm's total fixed costs. This allocation will be based on some accounting convention, but there is no 'correct' answer, so different people may arrive at different allocations of fixed costs. Such different levels will result in different break-even levels.

For example, if a company makes two products (X and Y) and employs five people and we know that the contribution from X is £10 and from Y is £20, with a total fixed cost of £10 000, then we may try to find break-even production levels for X and Y. If we were to allocate the total fixed costs on the basis of the labour involved in each product, we would get:

for X, $\dfrac{3}{5} \times$ £10 000 = £6000, and dividing by the contribution

(£6000/£10) gives a break-even production level of 600.

for Y, $\dfrac{2}{5} \times$ £10 000 = £4000, and dividing by the contribution

(£4000/£20) gives a break-even production level of 200.

However, if each product is produced in half of the available space, and we allocate fixed costs on the basis of floor space used, each product will be allocated £5000. This gives break-even production levels of:

for X, (£5000/£10) = 500, and
for Y, (£5000/£20) = 250

3.3   LINEAR PROGRAMMING

Linear programming, or LP, is a technique which aims to find the optimum allocation of scarce resources between competing products or activities. It is very widely used in a business context, and examples include:

optimum product mix
media selection
share portfolio selection

Where there are just two products or activities and few scarce resource constraints, then LP is a graphical technique which can find answers fairly quickly. You will need to specify the problem, or 'write the problem down', in such a way that a solution can be found. This is usually referred to as *formulating* the problem.

We are trying to create a graph which shows all feasible mixes of the products, media types or shares. We will limit our analysis to only two items, but you should note that the techniques will work in much more complex situations. In such cases you will need some software to help solve the problem, but if you have understood the two-variable problem, then it is not a particularly big step to move on to three, four or more variable problems.

A small company (Singletons & Co.) make two products.

They are asking for your advice on what mix of products to make, and have been able to provide the following information:

Ambers require **1** hour of labour time
Zeonites require **2** hours of labour time
Total labour hours per week is **40**

Ambers require **6** litres of moulding fluid
Zeonites require **5** litres of moulding fluid
Maximum moulding fluid per week is **150** litres

Profit contribution from Ambers is **£2**
Profit contribution from Zeonites is **£3**

How many of each product should Singletons & Co. make each week?

In order to solve this problem, we need to know what we are trying to do for this company. Although not specifically stated, the most likely thing they want is to make as much profit as possible, known as *maximizing profit*. Profit can only come from the two products they sell. So they make £2 for each Amber and £3 for each Zeonite sold. Using A and Z for the products, we could write the profit function as:

$$\text{Profit } (\Pi) = 2A + 3Z$$

If there were a plentiful supply of everything needed, then there would be no problem! We could just go on producing more and more, and making more and more profit. This is never the case! There are limitations on the amounts of Amber and Zeonite the company can produce.

The first restriction is in relation to labour. We know Ambers take 1 hour and Zeonites take 2 hours each. Therefore the total labour used is:

$$A + 2Z$$

But the maximum amount of labour available is 40 hours, so the total used must be equal to or less than this number. We have an **inequality** (see section 2.8 for details on how to graph an inequality):

$$A + 2Z \leq 40$$

Using the same argument for moulding fluid, we get:

$$6A + 5Z \leq 150$$

❗ *Check that this makes sense to you now!* The rest of this section will pass you by if you missed this point.

To understand what these inequalities mean, and to take the first steps in solving the problem, we will construct a graph. Taking the first inequality, let's look at the outer limit where $A + 2Z = 40$. This gives the graph shown in figure 3.7.

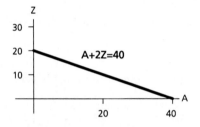

FIGURE 3.7   Graph of the labour restriction

Note that this is the outer limit, so anything below this line is feasible, until we get down to the axes. We can't go lower, since that would mean negative production. We usually shade this area, as in figure 3.8.

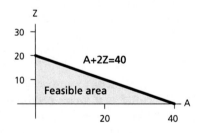

FIGURE 3.8   Feasible area for labour

Again we can use the same logic in relation to the moulding fluid restriction. We get the graph in figure 3.9.

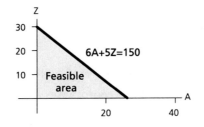

FIGURE 3.9   Feasible area for moulding fluid

Putting the two graphs together, and remembering that we cannot go past the axes, we get figure 3.10 which shows all of the feasible solutions to the problem.

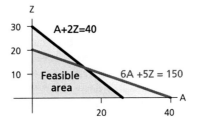

FIGURE 3.10   All feasible solutions

Any point in the **feasible area** will be possible, but remember that we were trying to maximize profits, so making more is viewed as better. This means that the solution will be as far away from the origin as possible – at one of the corners of the feasible area. There are several different ways of identifying which corner gives the 'best' solution, but for a two-variable problem with few restrictions or limitations, it is usually fairly quick to work out the profit at each corner and just pick the biggest.

Looking at figure 3.10, we could just read the values of A and Z from the graph and put them into the profit function. Alternatively, we could work out where the lines cross using algebra. Where the lines cross the axes it is fairly simple to read off the values (since one will be zero), but where the two limitations cross, algebra will help. The two lines are:

$$A + 2Z = 40$$
$$6A + 5Z = 150$$

and these are simultaneous equations which we can solve (see section 2.7 for details of solving pairs of equations).

Multiply the first by 6 and then subtract:

$$6A + 12Z = 240$$
$$\underline{6A + 5Z = 150}$$
$$7Z = 90$$
$$Z = 12.857$$

Substituting into one of the equations gives $A = 14.2857$

We now know the coordinates of the four corners of the feasible area:

$$(0,0); (0,20); (25,0); (14.29,12.86)$$

Putting these into the profit function gives us table 3.1:

(0,0)	(0,20)	(25,0)	(14.29,12.86)
0	60	50	67.14

TABLE 3.1   Putting the coordinates into the profit function

It is obvious which is the highest, so our recommendation to the company is to produce 14.29 Ambers and 12.86 Zeonites. Note that we would interpret these as average production figures per week since the actual number of products completed in a particular week would be integer (whole number) values: some incomplete products could be carried over to the next week if they are not perishable.

3.4   LINEAR PROGRAMMING – MINIMIZATION

This technique can also be used to find a minimum cost solution instead of a maximum profit one. The way you do this is basically the same, but this time you are trying to get as close to the origin as possible, rather than as far away as possible.

A company has two machines, A and B which can each produce either the MINI or MAXI version of their product:

A can produce **5** MINIs or **1** MAXI per session
B can produce **2** MINIs or **3** MAXIs per session

Contracts dictate that the minimum number:

of MINIs must be 100
of MAXIs must be 90

The cost of running machine A is **£1000** per session
The cost of running machine B is **£2000** per session

What is the minimum cost number of sessions for each machine?

Again, working through an example will illustrate the method.

The cost function is:

$$Cost = 1000A + 2000B$$

Looking at the number of MINIs produced, we have:

$$5A + 2B$$

The minimum we need is 100, so our inequality is:

$$5A + 2B \geq 100$$

By the same reasoning, for MAXIs we have:

$$A + 3B \geq 90$$

Again, we do not want to go below the axes, so:

$$A \geq 0 \text{ and } B \geq 0$$

Putting all of this information onto a graph and shading in the feasible area, we get figure 3.11.

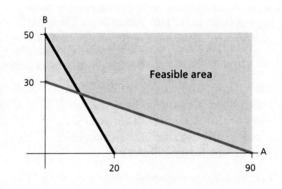

FIGURE 3.11   Feasible area for cost minimization

We find the coordinates of the corners of the feasible area as before. This gives:

$$(0,50); (90,0); (9.23,26.9)$$

Putting these values into the cost function gives us table 3.2:

(0,50)	(90,0)	(9.23,26.9)
£100 000	£90 000	£63 030

TABLE 3.2   Putting coordinates into the cost function

The lowest value is obvious. So we recommend to the company to use Machine *A* for 9.23 sessions and Machine *B* for 26.9 sessions. In practice, we would look for an integer solution.

Linear programming may seem too simple to be useful in practice, after all, we know that relationships are not necessarily linear. However, linear functions do prove to be particularly robust and work well in a very wide range of situations. The practical difficulties often revolve around specifying the problem, that is, finding the appropriate values to put into the equations, especially the contribution to profit for each product or service. Computer software is normally used to solve linear programming problems, especially when we have more than two products and many constraints. Such software can also often deal with situations such as requiring a whole number answer (known as integer programming).

3.5 | CONCLUSIONS

This chapter has been rather wide ranging, covering topics often found much later in typical courses. We have included them here since they show the practical application of the topics from Chapter 2.

Time series graphs are useful in giving initial impressions of the behaviour of data and are frequently used where only general trends are required.

Break-even analysis is fundamental to understanding whether a business or a business idea is worth working on. It is also useful if the organization needs to work within a budget. If there is no prospect of profit without massive sales and massive market share, then **payback** on the initial investment is likely to be in the distant future, if at all.

Linear programming provides a method of solution for a wide range of problems and is not limited to two items and a few constraints, as in our examples. Computer-based solutions are easily available, and for small problems you can use an add-in to Excel: see Jon Curwin and Roger Slater, *Quantitative Methods for Business Decisions* (Thomson Learning, 2002, Chapter 19) for more details. For larger problems there is specialist software available. Linear programming provides a short- to medium-term solution, but in the long run, managers need to address the resource constraints themselves if they wish to increase production levels.

When you have worked through and understood this chapter, you will be in a position to:

- use graphical methods in a variety of ways
- illustrate company and other data, e.g. sales figures, population trends
- advise on a break-even level for a single product company
- explain the concept of a feasible area and the way that this will constrain production or service levels

3.6 EXERCISES

1   Create a time series graph from the following data:

	Quarter			
Year	1	2	3	4
1	20	40	50	30
2	25	50	60	40
3	35	60	80	40

2   Create a time series graph from the following data:

	Quarter			
Year	1	2	3	4
1	10	50	30	10
2	20	100	60	20
3	40	200	130	50

3   If you have access to the Internet, go to the Statistics web site at **http://www.statistics.gov.uk** and download annual data on the percentage of people unemployed. Put this into a spreadsheet and construct a graph of the data. Comment on the graph. Experiment by constructing graphs of both quarterly and monthly data.

> ❗ Hint: you will need to construct the monthly one over only about three years, or the detail will be lost.

4   A company makes Sweetos and sells them at £10 per box. The costs of production consist of a fixed cost of £25 000 and a variable cost per box of £5. What is the break-even production level for this company?

5   McArlen's make McArlen Cakes and these sell to the trade at £14 per 100. The cost of making 100 cakes is £10. McArlen's also face an annual fixed cost of £100 000. What is the annual break-even production level for the firm?

6   Yarmi's produce specialist fireworks which they sell at £100 each. Their fixed costs are £10 000 and the variable cost of a firework is £50. What is their break-even production level?

7   Singh's arrange trips and have a fixed cost of £1000 per week. Their variable costs per trip can be represented by the function $VC = X - 20$ where X is the number of trips arranged. The price of a trip is £50. What is the break-even number of trips per week for this company?

8 Williams' Coaches hire out their largest coaches at £500 per day. Their total variable costs can be represented by the function $TVC = X^2 - 200X$ and their fixed costs are £60 000 per year. Within what range can they break even?

9 A company makes two products. It takes 20 minutes of labour time for Xeeny and 40 minutes for Yoomy. Total labour time available is 800 minutes per day. It also needs finishing time on a machine of 20 minutes for Xeeny and 80 minutes for Yoomy. Total machine time is 1200 minutes per day. The profit contribution of Xeeny is £4 and of Yoomy is £7. There is an agreement that the minimum level of production for Yoomy is 5. What are the profit maximizing levels of production for each product?

10 A manufacturer of fitted kitchens produces two units, a base unit and a cabinet unit. The base unit requires 90 minutes in the production department and 30 minutes in the assembly department. The cabinet unit requires 30 minutes and 60 minutes respectively in these departments. Each day 21 hours are available in the production department and 18 hours are available in the assembly department. It has already been agreed that no more than 15 cabinet units are to be produced each day. If base units make a contribution of £20 per unit and cabinet units make a contribution of £50 per unit, what production mix will maximize profit? What is this maximum profit?

11 A company has two lorries which it uses for deliveries. The first (X) can carry 10 of product A or 4 of product B. The second (Y) can carry 3 of A or 5 of B. Minimum deliveries are 200 of A and 150 of B. In order to maintain roadworthiness, each lorry must be used for a minimum of two journeys per week. If the running costs are £200 per journey for lorry X and £150 per journey for lorry Y, find the number of deliveries made by each lorry to minimize costs.

✔

3.7 ANNOTATED ANSWERS

1 If you draw the graph by hand, then you can work directly from the data. However, if you are going to use a spreadsheet, you need to transfer the data into a single column. You can then use the Chart Wizard to draw a line graph of this column. To make the graph easier to read, make sure that you fill in the section for x-axis values. This screen shot illustrates the data and the graph.

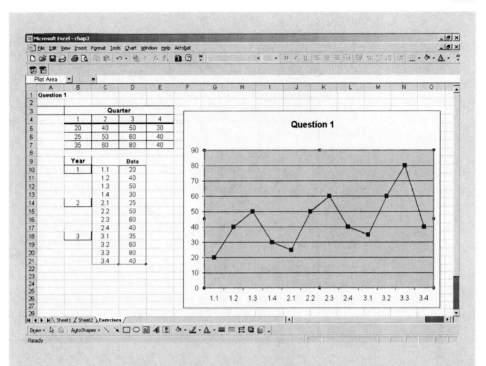

2   As in the previous question, we can transfer the data into a column of a
    spreadsheet and create a line graph:

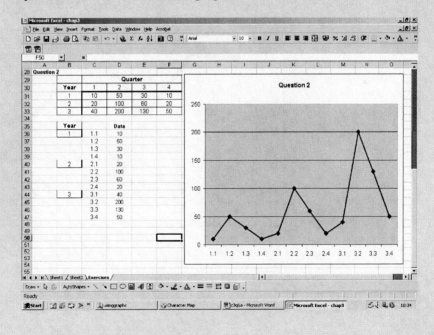

3 Your answer here will depend on when you do the question, and therefore the data available on the web site. As an example, when we were writing the book we got these answers:

Annual data:

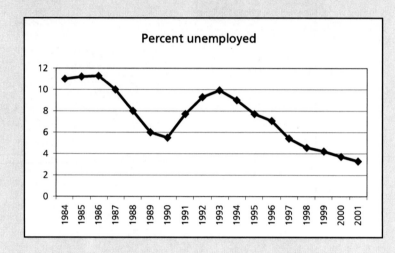

Quarterly data:

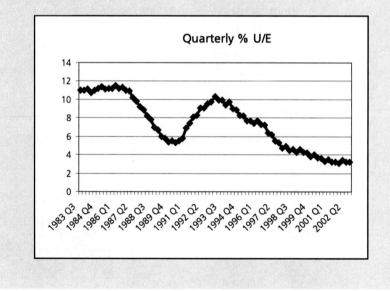

Monthly data:

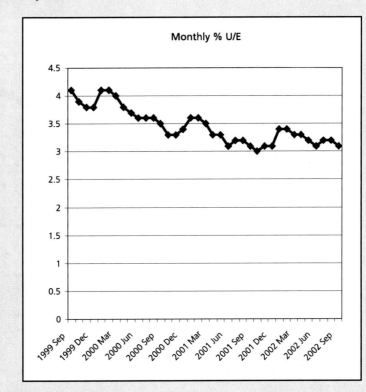

4   If the price per box of Sweetos is £10 and the variable cost is £5, then the contribution per box is (£10 – £5) = £5. Break even is defined as fixed cost divided by contribution, so we have:

$$£25\ 000/£5 = 5000$$

5   Using X to represent 100 cakes, the total revenue for McArlen's is:

$$TR = 14X$$

And the total cost is:

$$TC = 10X + 100\ 000$$

Drawing the graphs gives:

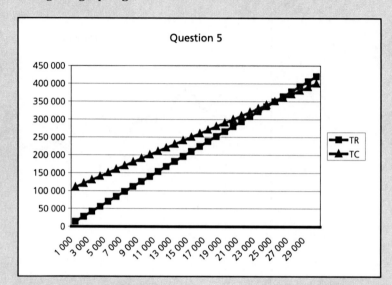

and you can read off the break-even value of 25 000 batches of 100 cakes. Alternatively, you can calculate the contribution as £14 − £10 = £4 and then divide this into the fixed cost of £100 000 to get the answer of 25 000 batches of 100 McArlen's cakes.

6 If X represents a firework, then $TR = 100X$ and $TC = 10\,000 + 50X$. Since profit is $TR - TC$, we have $= 100X - 10\,000 - 50X = -10\,000 + 50X$. This can be graphed to give:

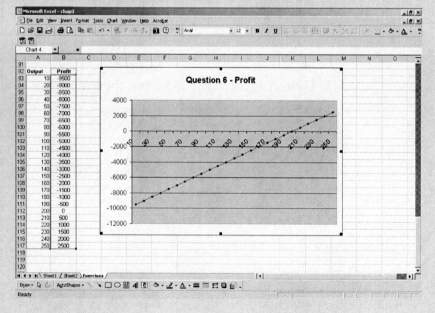

You can now read off the break-even level as 200 fireworks.

Alternatively, you can calculate contribution as £100 – £50 = £50 and divide this into the fixed cost to give (10 000/50) = 200 fireworks.

7   Here we have a non-linear cost function. The total variable cost is X times the variable cost of a trip, giving $X^2 - 20X$, so the total cost for Singh's is:

$$TC = X^2 - 20X + 1000$$

Total revenue is the price times the number of trips:

$$TR = 50X$$

Profit is $TR - TC$:

$$\Pi = 50X - (X^2 - 20X + 1000) = -X^2 + 70X - 1000$$

Break even is where profit is zero, so we could solve this as a quadratic equation, or we could draw the graph.

Using the quadratic formula, we get:

$$\text{Roots} = \frac{-70 - \sqrt{70^2 - 4^2(-1)^2(-100)}}{-2} = \frac{-70 - \sqrt{4900 - 4000}}{-2} = \frac{-70 - \sqrt{900}}{-2}$$

$$= \frac{-70 - 30}{-2} = \frac{-100}{-2} \text{ and } \frac{-40}{-2}$$

So break even is at 20 and 50 trips per week. They need to sell within these boundaries.

Drawing the graphs gives:

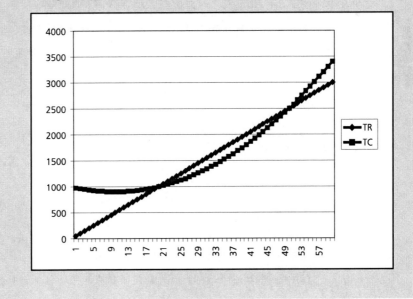

You can now read off the two break-even values of 20 and 50 trips per week.

8 From the information given, we get:

$$TR = 500X \text{ and } TC = X^2 - 200X + 60\ 000$$

Therefore profit will be:

$$\Pi = 500X - (X^2 - 200X + 60\ 000) = -X^2 + 700X - 60\ 000$$

Break even is where this is equal to zero, so we can use the quadratic formula:

$$\frac{-700 - \sqrt{490\ 000 - 4(-1)(-60\ 000)}}{-2} = \frac{-700 - \sqrt{250\ 000}}{-2} = \frac{-700 - 500}{-2} = \frac{-1200}{-2} \text{ and } \frac{-200}{-2}$$

giving answers of 100 and 600. So Williams' Coaches need to hire out between 100 and 600 coach days per year to break even.

9 As with any linear programming problem, the first step is to formulate it – change the words into algebra. Decide first on a single letter to represent each product – here X and Y seem obvious. Looking at the labour time constraint first, X takes 20, Y takes 40 and the maximum available is 800. This gives:

$$20X + 40Y \leq 800$$

Now we can look at the machine time constraint. Here the maximum is 1200 and X takes 20 while Y takes 80. This gives:

$$20X + 80Y \leq 1200$$

The last sentence tells us that we must make at least 5 of Y, so:

$$Y \geq 5$$

We also know that production cannot be negative, so:

$$X, Y \geq 0$$

Finally, profit contributions are given for each product so that we can create an expression for total profit (Z) as:

$$Z = 4X + 7Y$$

We have now formulated the problem!

To solve the problem we need to construct a graph. If we take each constraint in turn, use = instead of ≤ and find where it crosses the axes, firstly by letting $X = 0$, then by letting $Y = 0$, we will be able to do this.

For the labour constraint, if $X = 0$, $Y = 20$; if $Y = 0$, $X = 40$

For the machine time constraint, if $X = 0$, $Y = 15$; if $Y = 0$, $X = 60$

For the minimum production constraint, just let $Y = 5$ for all X values.

Putting these on to a graph gives us:

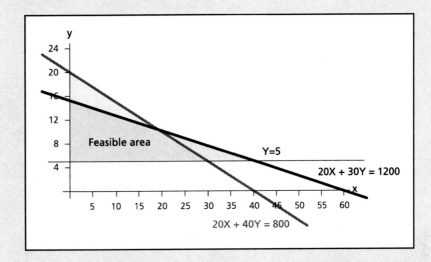

We can now find the X and Y vales at the corner points of the feasible area and evaluate the profit function at each one.

Corner	$X = 0, Y = 15$	$X = 20, Y = 10$	$X = 30, Y = 5$	$X = 0, Y = 5$
Profit	105	150	155	35

So maximum profit is made from production of 30 Xeeny and 5 Yoomy per day.

10 Again we have a maximization problem using linear programming.

We will use B for base units and C for cabinets, so now we can do the formulation.

First look at the production department: maximum hours are 21, B takes 1.5 hours and C takes 0.5 hours, so the constraint is:

$$1.5B + 0.5C \leq 21$$

For the assembly department, max is 18, B takes 0.5 hours and C takes 1 hour, so constraint is:

$$0.5B + C \leq 18$$

For cabinets there is a maximum production, so:

$$C \leq 15$$

We cannot have negative production, so:

$$B, C \geq 0$$

Profit is:

$$Z = 20B + 50C$$

To solve the problem we need to construct the graph to find the feasible area.

For the time constraint, if $B = 0$, then $C = 42$ and if $C = 0$, $B = 14$

For the assembly constraint, if $B = 0$ then $C = 18$, if $C = 0$ then $B = 36$

There is also the maximum production of C constraint

Putting these onto a graph gives:

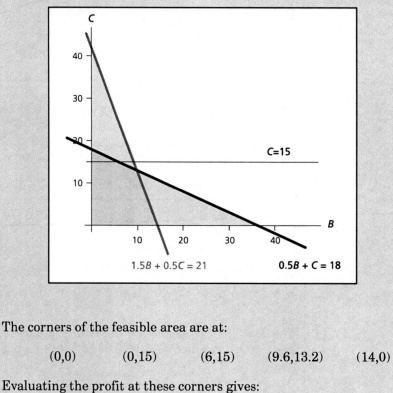

The corners of the feasible area are at:

$(0,0)$ $(0,15)$ $(6,15)$ $(9.6,13.2)$ $(14,0)$

Evaluating the profit at these corners gives:

0   750   870   852   280

So we recommend that the company makes 6 base units and 15 cabinet units per week.

11 This is a cost minimization problem, so we need to minimize the cost function:

$$200X + 150Y$$

The limitations are given for each product (A and B):

For A	$10X + 3Y \geq 200$
For B	$4X + 5Y \geq 150$
We also have:	$X, Y \geq 2$

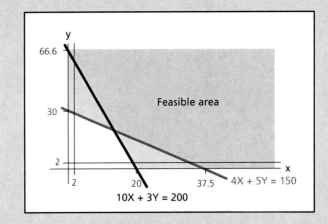

By making the inequalities into equations, we can find where they cross the X and Y axes. Anywhere above the constraints is feasible, so we have three points to evaluate using the cost function: $200X + 150Y$

These are (2,60), (14.47,18.42) and (35,2) and give costs of 9400, 5657, and 7300. The minimum is 5657, so the optimum point is at 14.47 journeys by lorry X and 18.42 journeys by lorry Y. (Since these are not whole numbers, we could treat them as the average number of journeys per week.)

WEB REFERENCE 👁

www.thomsonlearning.co.uk/businessandmanagement/curwin3

See the companion web site for further questions and annotated answers. There you will also find a PowerPoint presentation which you can use to help understand this area, or later, for revision. The site also contains links to other sites of interest.

Chapter 4

REPRESENTING DATA

A very old saying has it that 'a picture is worth a thousand words' and we find that using graphs (or pictures) is particularly helpful in communicating statistical ideas. Unfortunately, it is just as easy to misuse such graphs, or use inappropriate graphs. Reading this chapter will help you choose appropriate graphs and diagrams which should be informative and of interest.

OBJECTIVES

After reading this chapter and doing the exercises you should be able to:

- construct bar charts
- construct pie charts
- construct pictograms
- construct histograms
- argue for or against a particular statistical graph in the context of the data
- use Excel to produce such graphs

4.1 | WHY USE STATISTICAL GRAPHS?

The primary aim of a graph in this context is to communicate numerical information to your audience. Many people, even some in managerial roles, are uncomfortable with figures but we still need to present numerical ideas to them, maybe by using PowerPoint. A graph is able to do this, if used carefully. You might be asking if statistical graphs are, in any way, different from the graphs dealt with in Chapters 2 and 3. The answer is 'yes'. The types of mathematical graphs dealt with earlier are used to illustrate algebraic functions, such as a linear function, whilst these graphs and diagrams aim to illustrate real world data. Where there is a gap in a mathematical graph you can often deduce what the missing values are by using algebra. In the case of these graphs, a missing figure can only be guessed at. It is true that you might infer a trend from the data, but the actual data for that particular case cannot be deduced from the graph.

Whilst you may have to draw simple charts by hand, it is usual, and more convenient, to use Excel, or a similar package, to create these statistical graphs. This mostly makes your life easier, but there are a few things to be careful of as you specify what exactly it is that you want graphed. See the section at the end of the book and on the web site for more help in using Excel.

4.2 | SIMPLE BAR CHART

This is probably the most widely used graph because it is easy to draw and easy to understand – the longer the bars in the chart, the bigger the quantity they represent. Simple **bar charts** can be drawn with the bars vertical or horizontal, depending on your own preference.

Suppose that you work for a cosmetics manufacturer who wants to launch a range of products keyed to the person's eye colour. You have commissioned a survey which has determined the following data:

Eye colour	Number of people
Green	60
Blue	40
Brown	100

Construct a bar chart for this data.

If we put the bars horizontally, one for each eye colour, we will need a scale on the *x*-axis, which goes from 0 to 100. We can then make each bar the same width and their length is measured on this horizontal scale as in figure 4.1.

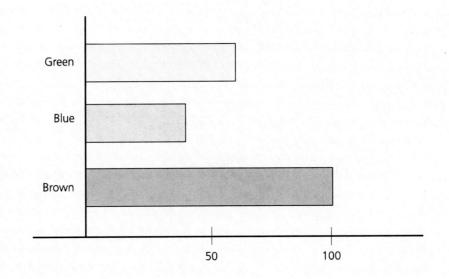

FIGURE 4.1 Simple bar chart

A bar chart is particularly good for **categorical data** as you can have one bar for each category. For this reason they are often used in reports on the results of questionnaires (see Chapter 12) where many of the questions have **pre-coded answers**. Other frequent uses are in company reports and reports by local councils.

Most bar charts will be created on a spreadsheet. Excel will create the straightforward bars we see above, but will also create tubes, three-dimensional bars, and many other variations. We show the basic bar chart, figure 4.2, tubes, figure 4.3, and cones, figure 4.4, below.

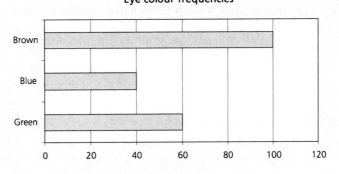

FIGURE 4.2 Simple bar chart produced by Excel

Number of people

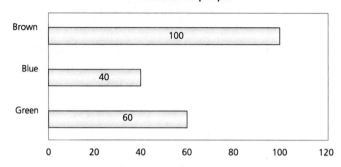

FIGURE 4.3   Tubes bar chart as produced by Excel in Custom Types of chart

Number of people

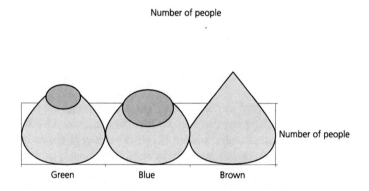

FIGURE 4.4   Cones bar chart as produced by Excel in Custom Types of chart

However we need to be careful, as the form of comparison becomes more complex, that we don't lose sight of the relative magnitudes involved.

 Would you say the numbers involved (60, 40 and 100) are as clear in figure 4.4 as in figure 4.1?

4.3 │ MORE COMPLEX BAR CHARTS

A bar chart is capable of showing rather more information than that shown in the previous section. Each of the bars can be sub-divided and then shaded or colour coded. The bars can be constructed in exactly the same way as those in the previous section, or they can all be made the same length and the sub-divisions then show percentage comparisons. Again we will work through an example to illustrate the construction of these charts.

Suppose we have the results of a European survey asking about the numbers of men and women who are involved in a minority sport. These results are shown below:

	Men	Women	Total
British	40	30	70
French	75	15	90
German	60	40	100
Italian	40	70	110
Dutch	20	100	120
Greek	90	20	110

To illustrate this data we want to draw a **component bar chart**.

Each bar will represent a nationality. The length of the bar will represent the total number of people, but each bar will be sub-divided into the number of men and the number of women from that country. If you have the data on a spreadsheet, highlight all of the data for nationality, men and women, but not the total. You can then click on the Chart Wizard icon, 📊 on the tool bar, and select Bar Chart in the new window which appears. From the range on the right, select the Stacked Bar option and click your way through the various steps of the wizard. Your graph should look something like figure 4.5:

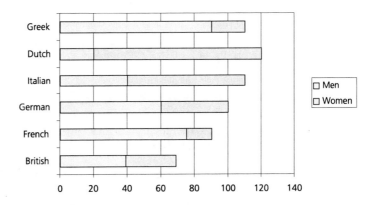

FIGURE 4.5   Stacked bar chart

This sort of bar chart is relatively easy to understand when there are just two divisions to each bar, but if you had a larger number, say 10 divisions, it would be increasingly difficult to understand. Even with just two sections, it can be quite difficult to compare the relative number of women between Britain and Germany from the chart.

An alternative is to make all of the bars the same length and convert the numbers to percentages – this makes comparisons between countries much easier. This is very time consuming to do by hand, but by using a spreadsheet like Excel, it can be done in seconds. Follow the steps as before, but this time, select the 100 per cent stacked bar option from the types of bar chart. Your graph should look like figure 4.6.

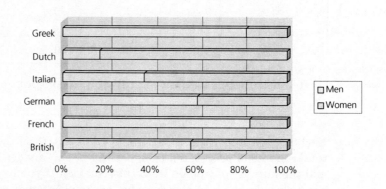

FIGURE 4.6   100 per cent stacked bar chart

Now comparisons are easier, but if there were large differences in the actual number of people from each country, then the graph might be very misleading. These representations are also known as component bar charts.

4.4   PIE CHARTS

A **pie chart** also offers a quick and easy way to illustrate the results of a survey question. This is one chart you may recognize from school. Again, it is a graph which could take considerable effort to produce by hand, but which can very quickly be generated using a spreadsheet.

The basic principle of a pie chart is that the total amount which we are trying to illustrate is represented by the 360 degrees of the circle. The various sub-groups in this total have a 'slice of the pie' which is proportional to their size – a proportion of the 360 degrees. We can then construct the pie chart such that all of the slices together make a circle.

If we take the previous example on eye colour, we have table 4.1:

Eye colour	Number of people
Green	60
Blue	40
Brown	100

TABLE 4.1   Eye colour table

The total number of people is 200, so one person is represented by (360/200) = 1.8 degrees of the circle. Applying this to the table, would give us the results in table 4.2:

Eye colour	Number of people	Degrees
Green	60	108
Blue	40	72
Brown	100	180

TABLE 4.2   Eye colour table with degrees of circle applied

From this we could draw the diagram using a protractor.

Alternatively, we could put the data into a spreadsheet and produce a pie chart, using the Chart Wizard. A simple pie chart from Excel is shown in figure 4.7.

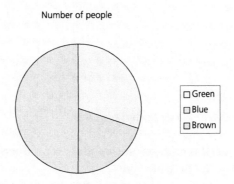

FIGURE 4.7   A pie chart

As with the bar chart, the spreadsheet offers a variety of pie charts, some in three dimensions. Two versions are illustrated below (figures 4.8 and 4.9).

Here the exploded pie chart produced by Excel has been edited by changing the colours to match the labels and by moving the slices so that only one of them is separated from the rest. This gives rather more emphasis to one sub-group, here those with blue eyes.

Number of people

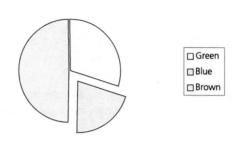

FIGURE 4.8   An exploded pie chart

 Note that you can edit an Excel picture by double-clicking on the area you want to edit.

In figure 4.9 the standard Excel output has been modified by changing the view angle. This is done by using the Chart menu in Excel.

Number of people

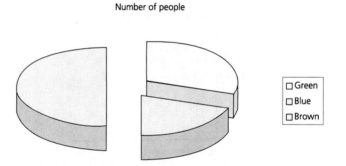

FIGURE 4.9   3D exploded pie chart

Using the data on nationality, reproduced below in table 4.3, we can make a more complex pie chart.

	Total	Men	Women
British	70	40	30
French	90	75	15
German	100	60	40
Italian	110	40	70
Dutch	120	20	100
Greek	110	90	20

TABLE 4.3   Data on nationality

The main chart shows the distribution of the total number of people by national-ity, and we have then highlighted the British position by drawing a separate pie chart (figure 4.10) to show the gender balance in Britain.

❗ Note that the extra label and arrow had to be added to the chart after it was drawn.

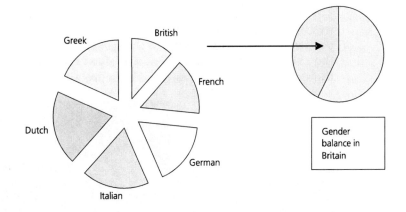

FIGURE 4.10   Using two pie charts together

Pie charts work well provided that there are up to about seven categories (a rule of thumb). More than this, and it becomes increasingly difficult to distinguish which group is which, and the relative size of the groups. This is illustrated in figure 4.11.

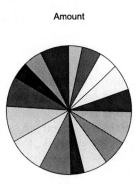

FIGURE 4.11   Too many sub-groups for clear representation by a pie chart

Again, pie charts are often used in company reports and reports by local coun-cils, for example in a council tax demand.

4.5 | PICTOGRAMS

Where we want to create an impact with our diagrams, rather than just illustrate the data, we might use pictures which are related to the subject matter. These pictures can then vary in size in proportion to the data we have. Such pictograms are often used in newspapers and in materials sent out to the general public, rather than in research reports.

The difficulty with such pictograms is that there is not necessarily any standard way of linking the size of the picture to the data. For example, the height of the picture could be proportional to the size of the data, or it could be the volume of the diagram. These two choices will give very different diagrams from exactly the same data. This is illustrated in figure 4.12.

Doubling in height:

Doubling in volume:

© Roger Slater

FIGURE 4.12   Pictograms representing a doubling of sales using height and volume

In this diagram, the first pair of pictures seem to show a much larger increase than the second pair. Many people using such pictures neglect to show a key, and therefore it is impossible to make a valid comparison. **Pictograms** often succeed in grabbing attention, and if that is your objective then they are a suitable method of illustration. For more academic reports they should be avoided.

4.6 | HISTOGRAMS

Histograms are the method of illustration 'preferred' by statisticians since they form a link with the development of theory and between discrete and continuous data. They can do this because in a histogram the area of a block represents the appropriate frequency. Histograms bear a passing resemblance to column bar charts, but in a bar chart it is only the length (height) which represents the frequency, not the area.

Where we have **discrete data**, that is, where there are only whole number answers, then a histogram looks like a bar chart. (Many people would not call this a histogram since the discrete data means each block is the same width, but it is a matter of interpretation.) For example, if we asked about the number of children in a family, there are only whole number answers, so our histogram would consist of blocks all of equal width, whose height was related to the frequencies as shown below in table 4.4.

Number of children	Number of families
0	40
1	30
2	20
3	8
4	2

TABLE 4.4   The number of children in a family

This table allows us to create the histogram shown in figure 4.13. Here the blocks are butted up against each other, but since they are of equal width, the heights are proportional to the frequencies.

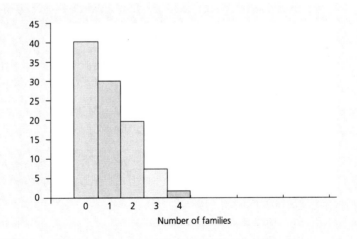

FIGURE 4.13   Histogram using discrete data

Of much more use is a histogram which represents **continuous data**, that is, where the variable can take on any value. Some variables which are continuous, for example, age, are often collected as 'age at last birthday' and thus become presented as discrete. Often such continuous data is grouped for presentational purposes, or is the result of a survey questionnaire where people have been asked to place themselves into a category such as an income band. These groups are not necessarily of equal width, and therefore the blocks which we will use to represent them will also be of unequal widths. Since they are of unequal widths and we are using area to represent frequencies, we need to adjust height to make area proportionate to frequency.

As an example, suppose that a questionnaire had included a question on income where respondents were asked to indicate which group they belonged to, and these were the results:

Income group	Number of people
£5000 but under £10 000	50
£10 000 but under £15 000	80
£15 000 but under £20 000	100
£20 000 but under £30 000	80
£30 000 but under £50 000	40
£50 000 but under £100 000	25

We want to represent this data by a **histogram**.

Excel does not yet produce a histogram where the blocks need to be of different widths, so we must revert to drawing by hand, usually on to graph paper!

The first step is to choose one of the groups to be a 'standard width'. All this means is that we will relate the widths of all of the other groups back to this one. The is no correct selection, just the one you feel comfortable with. In our case we will pick a width of £5000 as our standard; this is because three of the groups are this width and it will therefore involve fewer calculations. Looking at all of the groups we get the results in table 4.5:

Income group	Relation to standard width	
£5000 but under £10 000	Standard	1
£10 000 but under £15 000	Standard	1
£15 000 but under £20 000	Standard	1
£20 000 but under £30 000	Twice as wide	2
£30 000 but under £50 000	4 times as wide	4
£50 000 but under £100 000	10 times as wide	10

TABLE 4.5   Income by group

To get the heights of the block that we are going to draw, we divide the frequencies by the number representing the width. This gives us table 4.6:

Income group	Relation to standard width		Frequency	Height
£5000 but under £10 000	Standard	1	50	50
£10 000 but under £15 000	Standard	1	80	80
£15 000 but under £20 000	Standard	1	100	100
£20 000 but under £30 000	Twice as wide	2	80	40
£30 000 but under £50 000	4 times as wide	4	40	10
£50 000 but under £100 000	10 times as wide	10	25	2.5

TABLE 4.6   Income by group with frequency and height added

We are now in a position to construct the histogram. We need a linear scale on the x-axis and then we draw the blocks onto this. This is shown in figure 4.14.

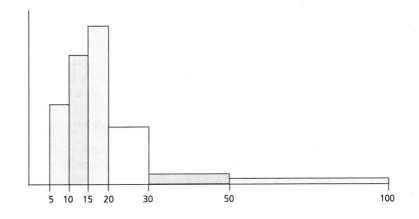

FIGURE 4.14   Histogram with continuous data

Sometimes the midpoints of each group are joined together to create a *frequency polygon* which may help you to see the overall shape of the distribution. If you take this subject further, then you might compare the frequency polygon to some standard shapes of distributions and draw more implications from this. A frequency polygon is shown in figure 4.15.

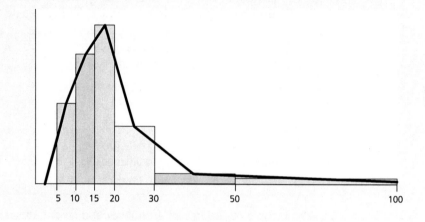

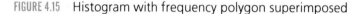

FIGURE 4.15   Histogram with frequency polygon superimposed

4.7 | USING EXCEL

As you will have seen if you have been using Excel to create the diagrams in this chapter, there are a wide variety of statistical charts available to you at the click of a mouse button. Whilst some of these may be useful to add variety to your presentations and reports, the use of too many within one report can be off-putting. If your audience is unfamiliar with some of the chart types you choose, then you may fail to communicate – the first objective of using charts and diagrams.

The ability to quickly produce clear diagrams in three dimensions from data has transformed many presentations. As a final illustration in figure 4.16, we have produced a three dimensional column chart to compare two groups of men and women:

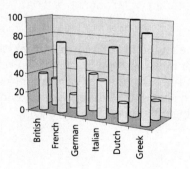

FIGURE 4.16   3D chart

Further details on Excel are given in the final section of the book, and also on the web site.

4.8 | CONCLUSIONS

Charts are all about communication, and this should be the way in which you judge your own and other people's efforts. The key question must be *Does it tell the correct story of the data?* If the answer is *Yes*, then it is an appropriate diagram. The chapter has looked at many diagrams, but it is up to you to choose appropriate ones in practice. Try to avoid too much variety as this becomes tedious to the reader, but similarly, don't just use a simple bar chart in every case.

When you have worked through and understood this chapter, you will be in a position to:

- represent data from a variety of sources appropriately
- justify your choice of diagram
- be aware of the use of Excel in this context

4.9 EXERCISES

1   The results of a questionnaire show the following data:

Question: 'How many children do you have?'

Answers:

No. of children	No. of answers
0	40
1	30
2	20
3	8
More than 3	2

Present this data using a suitable diagram.

2   Use a spreadsheet to create at least three different bar charts from the following data:

No. of times per week	No. of answers
0	30
1	10
2	15
3	30
4	50
5	80
6	60
7	40
More than 7	10

3   A shop is interested in the number of customers in the mornings and afternoons and has collected numbers from the previous week. You have been asked to prepare at least two stacked bar charts to illustrate the data and to comment on the advantages and disadvantages of each.

Day of the week	Morning	Afternoon
Monday	20	35
Tuesday	40	45
Wednesday	45	55
Thursday	50	60
Friday	60	90
Saturday	80	120
Sunday	20	10

4   Construct at least two types of pie chart to represent the following data:

Students in a residence block	
Degree subject	Numbers
Business	50
Nursing	80
Computer Science	70
Sociology	20

5   A small, local company monitors the area where it gets orders from and has collected the following data:

Area	Number of orders
Birmingham	50
Walsall	40
Wolverhampton	20
Dudley	35
Halesowen	40

Illustrate this data by a simple pie chart and by your choice of three-dimensional pie chart.

6   Many professions have a period of training and examination before you are allowed to practice. The table below shows the numbers starting training and those who subsequently both practise or drop out for a range of professions in a certain area.

Profession	Number entering	Into practice	Drop outs
Lawyer	400	100	300
Accountant	600	120	480
Nurse	1000	100	900
Doctor	300	100	200

Construct a pie chart of the total number entering each profession and select one of these to have a linked, subsidiary pie chart showing the numbers practising and dropping out.

7   Use a pictogram to illustrate the following data:

Sports club memberships	Number of people last year	Number of people this year
Golf	560	600
Tennis	400	320
Fitness gyms	300	900
War-gaming	30	30

Explain how you have chosen your pictures and the method of scaling used. How far does your pictogram succeed in its purpose?

8 Use a series of pictograms to represent this time series:

Year	1	2	3	4
Wine production (millions of litres)	45	55	75	80

9 Construct a histogram from the following continuous data:

Group	Frequency
0 but under 10	20
10 but under 20	30
20 but under 40	30
40 but under 80	10

10 A questionnaire included a question on how long each week the respondent watched television. The results have been grouped together into a table which is shown here. Use a histogram to illustrate this situation and then superimpose a frequency polygon onto your answer.

Time spent watching television	Number of people
0 but under 2 hours	20
2 but under 6 hours	80
6 but under 15 hours	200
15 but under 30 hours	120
30 but under 50 hours	40
50 but under 100 hours	10

1   We need five bars in the answer, one for each category. We also need a scale which goes up to 40 on the other axis. Just make each bar the same width, but the length measured on the 0 to 40 scale.

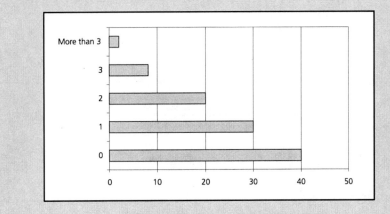

2   Make sure that you have highlighted the numbers and titles that you want to use for the bar chart. Then click on the Chart Wizard icon 📖 on the tool bar at the top. You can then select the type of chart that you want. A simple bar chart would be:

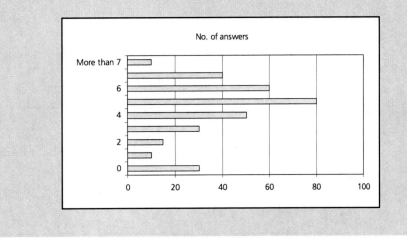

A tubes bar chart might be like this:

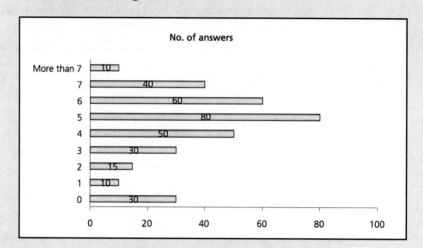

And a cylinder bar chart like this:

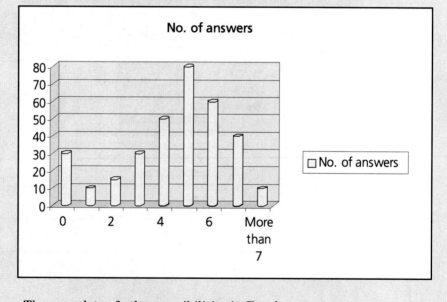

There are lots of other possibilities in Excel.

3 If you just select a stacked bar, with 3D effects, you should get:

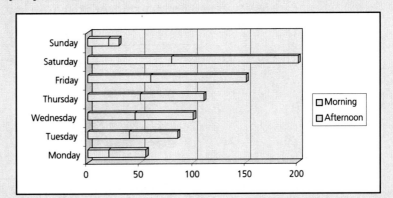

But you could select the column version with a 100 per cent stacked bar and 3D effects to give you this graph:

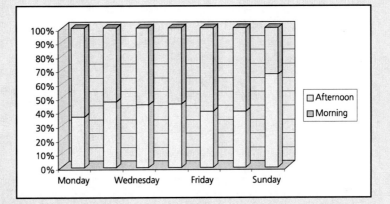

4 Using Excel there are several charts which can be produced. Firstly we have the basic pie chart:

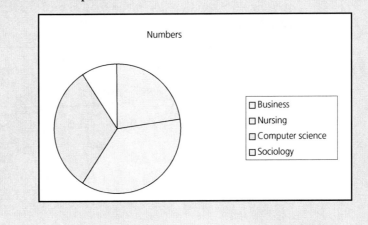

Secondly, we have used a 3D version but have taken the option to put data labels on to the slices, rather than have a key, or legend, by the side of the pie chart.

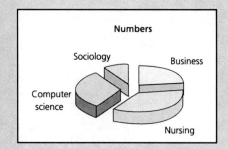

5   The simple pie chart is found by clicking on the chart icon and selecting the first pie:

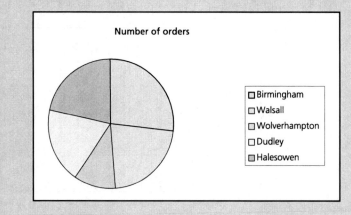

Our second pie chart uses a 3D view and has both labels and percentages attached to each slice. We have also pulled out one slice, Walsall, to add emphasis. This slice was moved by editing the picture and selecting to one slice. By holding down the mouse button, the slice can be dragged out of the pie.

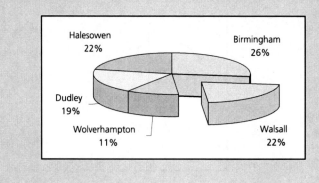

6 The main pie chart is constructed as usual. We then draw another chart just for the profession selected here, 'lawyer' and put the charts side by side.

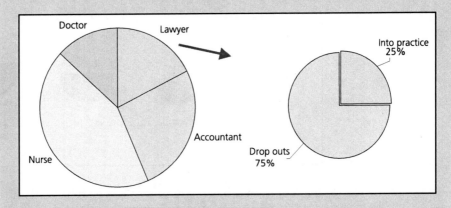

7 Pictures could be taken from books or magazines by scanning, photocopying, downloading from the net, or drawn by hand (depending on your talent). You could use a picture of a person throughout, or you could select pictures of each of the different 'sports' – this latter will make comparisons even more difficult.

8 In this case it seems obvious to select either bottles or glasses of wine as our pictogram. The difficulty is getting them to scale, but if you use height as your scaling factor, then the result is similar to a bar chart.

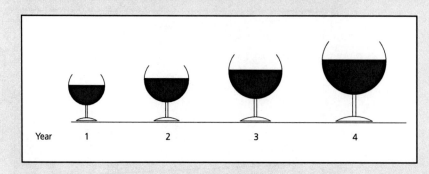

9 Remember that in a histogram that area represents frequency. We need to select one width as our standard – here we could use 10 as the standard width and get this histogram:

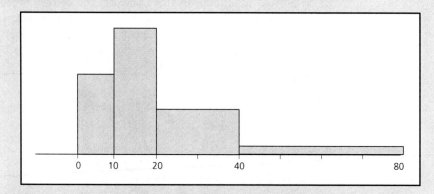

10 Again we select one width as our 'standard' and relate all of the others to that one. The problem here is that each group is a different width. We took the first one as standard and related each of the others to that one. This gave the table:

Time spent watching television	Number of people	Width	Relative width	Height
0 but under 2 hours	20	2	1	20
2 but under 6 hours	80	4	2	40
6 but under 15 hours	200	9	4.5	44.4
15 but under 30 hours	120	15	7.5	16
30 but under 50 hours	40	20	10	4
50 but under 100 hours	10	50	25	0.4

And the histogram:

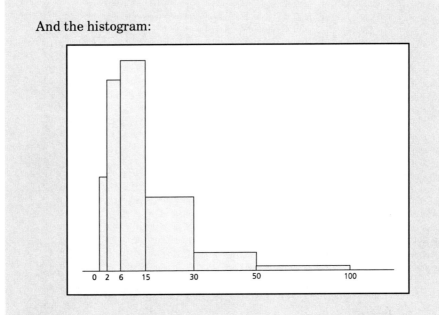

WEB REFERENCE 👁
www.thomsonlearning.co.uk/businessandmanagement/curwin3

See the companion web site for further questions and annotated answers.
There you will also find a PowerPoint presentation which you can use to
help understand this area, or later, for revision. The site also contains links
to other sites of interest.

Chapter 5

AVERAGING

It is one thing collecting lots of data but another actually making sense of it. Data provides a means to an end. It should allow us to address the purpose of our research. Rather than focus on every value, we are likely to become concerned with what is typical. We might ask, for example, what the average cost of private motoring is or the average level of credit card debt. The 'average' does have an everyday meaning and there is an awareness that an average can be found by adding values and dividing by the number of values included.

In this chapter we will clarify how we can measure the average and consider the correctness of calculation. Computer packages do make such calculations relatively simple but we still need to be aware of the assumptions being made and the limitations of such calculations. In particular, we will look at the mean, median and mode.

OBJECTIVES

After reading this chapter and working through the exercises you should be able to:

● describe data using a measure of average
● calculate the mean, median and mode for various types of data
● use a spreadsheet when appropriate
● assess the merits of using each of these measures of location

5.1 | CALCULATIONS FOR A SIMPLE LIST OF DATA

It could be the case that we merely have a list of numbers and need to calculate the mean, median and mode. Suppose we have been given the salary of five students who have recently been offered a one-year placement:

$$£12\,000 \qquad £9600 \qquad £12\,000 \qquad £10\,500 \qquad £13\,100$$

The mean

To calculate the **mean**, we add these five numbers together and divide the total by 5.

$$\bar{x} = \frac{12\,000 + 9600 + 12\,000 + 10\,500 + 13\,100}{5}$$

$$= \frac{57\,200}{5} = £11\,440$$

where $\bar{x}$ (pronounced 'x bar') is the symbol used to represent the mean.

In this case, the mean amount paid for a one-year placement is £11 440. It is interesting to note that no student is actually paid the average amount; two are paid less and three are paid more. The mean is a summary **statistic** and we need to be careful with interpretation. To say that the average household has 1.3 cars does not imply fractional ownership of cars but that many households have one car (1.3 is nearer to 1 than 2) but a number will own two or more.

To summarize the steps taken in calculation a shorthand, or notation, has developed. The calculation of the mean can be summarized as follows:

$$\bar{x} = \frac{\sum x}{n}$$

where x represents individual values,
$\sum$ (sigma) is an instruction to sum values
and n the number of values included.

The median

The **median** is the middle value of an ordered list. If we know the median age or the median income, for example, we would know that 50 per cent would be above this value and 50 per cent below.

The first step is to order the values:

$$£9600 \qquad £10\,500 \qquad £12\,000 \qquad £12\,000 \qquad £13\,100$$

We then need to count from the left (or the right). When working with a listing of values (but not with continuous data) the position of the middle value is found by using the formula:

$$(n + 1)/2$$

In this case we are looking at the 3rd position: $(5 + 1)/2$. The median is £12 000.

When working with an even number such as 6, a ½ emerges and we need to average the two adjacent values. Suppose we included an additional student in this data and this student had been offered a placement salary of £11 000:

| £9600 | £10 500 | £11 000 | £12 000 | £12 000 | £13 100 |

The formula $(6 + 1)/2$ gives 3½ and we need to average the 3rd and 4th values. The median is equal to $(11\ 000 + 12\ 000)/2$ or £11 500.

The mode

The **mode** is the most frequently occurring observation. We would use the mode if we were interested in the most frequent shoe size or most frequently bought model of car.

Given the data:

| £12 000 | £9600 | £12 000 | £10 500 | £13 100 |

the mode is £12 000 (occurs twice while the others occur once).

The determination of the mean, median and mode using Excel is shown in figure 5.1.

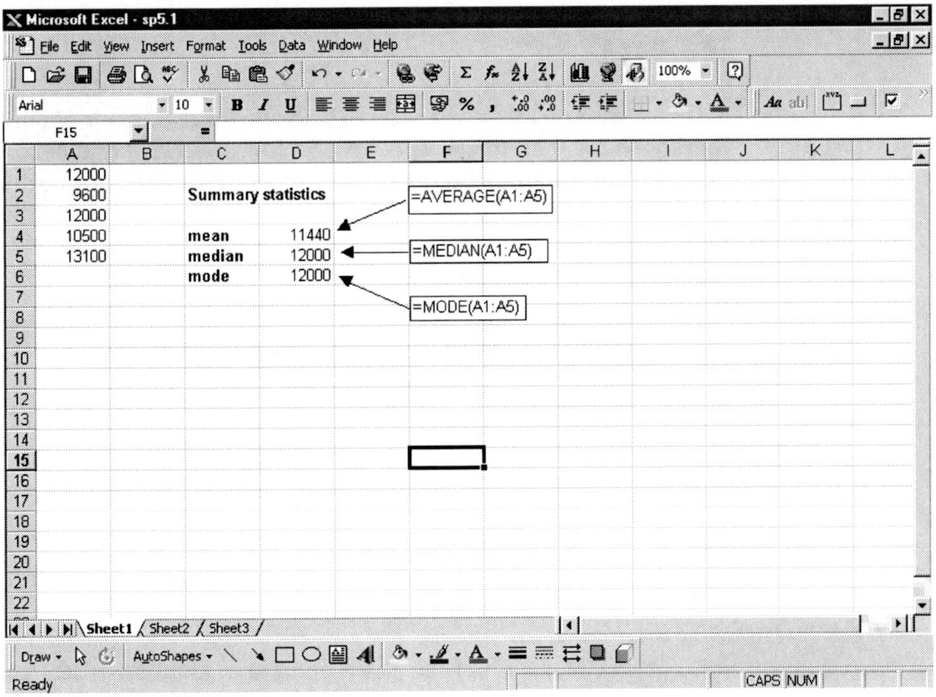

FIGURE 5.1   A spreadsheet showing the mean, median and mode

5.2 | CALCULATIONS FOR TABLES OF DISCRETE DATA

Results are often presented in the form of a table, e.g. results from a question-naire. The data given in the table 5.1 is referred to as **discrete data** because we are working only with whole numbers, e.g. the number of errors in a page of text or the number of children.

Number of children	Number of families
0	40
1	30
2	20
3	8
4	2

TABLE 5.1   Table of discrete data

This table summarizes the number of children recorded in 100 families (40 + 30 +20 + 8 + 2). In this case, we need to take account of the number of families with a given number of children. In this type of example, the number recorded (e.g. number of families) are usually referred to as the **frequency**, and the table referred to as a *frequency table*.

The mean

To calculate the mean we need to find the total number of children. The first 40 families have 0 children, the next 30 families have one each (30 children), the next 20 families have two each (40 children), the next 8 families have three each (24 children) and the next 2 families have four each (8 children). The total is found by multiplying the number of children by the corresponding frequency and then adding as shown in table 5.2.

Number of children (x)	Number of families (f)	fx
0	40	0
1	30	30
2	20	40
3	8	24
4	2	8
	100	102

TABLE 5.2   A calculation of the mean number of children in 100 families

The mean number of children in this sample of 100 families is 1.02.

The calculation of the mean in this case can be summarized as follows:

$$\bar{x} = \frac{\sum fx}{n} = \frac{102}{100} = 1.02$$

where f represents frequency.

The median

The table giving number of children from 0 to 4 has effectively ordered the data. Given that the total number of families is 100, we can use the formula $(n + 1)/2$ to find the position of the median (the table effectively provides a listing of forty 0s, thirty 1s and so on). The median corresponds to the $50\frac{1}{2}$th ordered observation. In this table the answer may be obvious. Given that 40 families have 0 children, and the next 30 have 1 child, the 50th and 51st will both have 1 child; the median is therefore 1 child.

In many tables the answer may be less obvious and working out **cumulative frequency** first is helpful. Cumulative frequency is the number of items with a given value or less. The determination of cumulative frequency is shown below in table 5.3:

Number of children	Number of families	Cumulative frequency
0	40	40
1	30	70
2	20	90
3	8	98
4	2	100

TABLE 5.3   The determination of cumulative frequency

It can be seen from the cumulative frequency column that both the 50th and 51st value is 1.

The mode

The mode corresponds to the highest frequency count. In this case, the mode is 0 children (the 40 families).

5.3 | CALCULATIONS FOR TABLES OF CONTINUOUS DATA

A summary table can also include a range of values in each row. Journey times could be given in 1 hour intervals or weight in 5 kilogram intervals, for example. We talk about **continuous data**, when a 'ruler' level of measurement has been achieved. You can measure something like time to any level of accuracy (providing a suitable measurement instrument exists). Time could be measured to the nearest hour, nearest minute, nearest second and so on. Income can be summarized in a table using equal intervals, or as shown in table 5.4, using unequal intervals.

Income group	Number of people
£5000 but under £10 000	50
£10 000 but under £15 000	80
£15 000 but under £20 000	100
£20 000 but under £30 000	80
£30 000 but under £50 000	40
£50 000 but under £100 000	25

TABLE 5.4   Income summarized using unequal intervals

In this case, we no longer know the exact income of any one person, we only know the range their income falls within. We know, for example, that 50 people have an income (or gave the response) '£5000 and under £10 000'.

The mean

To calculate the mean, we identify the **mid-point** for each income group and use this as our x value. Given the number involved, we can work with income in terms of £000s. The determination of the total number of people (n) and their total income ($\sum fx$) is as shown in table 5.5.

Income group (£000s)	Mid-point (x)	Number of people (f)	fx
5 but under 10	7.5	50	375
10 but under 15	12.5	80	1000
15 but under 20	17.5	100	1750
20 but under 30	22.5	80	1800
30 but under 50	40	40	1600
50 but under 100	75	25	1875
		375	8400

TABLE 5.5   Determination of the total number of people and their total income

The formula remains unchanged

$$\bar{x} = \frac{\sum fx}{n} = \frac{8400}{375} = 22.4$$

but x now represents mid-points.

The mean is 22.4 (£000s) or £22 400.

The median

We can determine the median either graphically or by calculation. In either case, we will need cumulative frequency and this is shown below in table 5.6.

Income group (£000s)	Number of people (f)	Cumulative frequency
5 but under 10	50	50
10 but under 15	80	130
15 but under 20	100	230
20 but under 30	80	310
30 but under 50	40	350
50 but under 100	25	375
	375	

TABLE 5.6   Income summarized showing cumulative frequency

The median – using the graphical method

To find the median graphically, we plot cumulative frequency against the upper boundary of the corresponding interval. In this case, the graph would show that 50 people had an income of £10 (thousand) or less, that 130 people had an income of £15 (thousand) or less and so on. The resultant graph is referred to as an **ogive**, and is shown as figure 5.2.

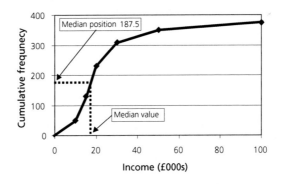

FIGURE 5.2   An ogive showing the determination of the median

To identify the median value, we use $n/2$ to identify position, not $(n + 1)/2$. The essential difference is that we are now dealing with a density (or distribution of continuous values) and not just a list of numbers. The median position is 187.5 (375/2) and this can be used to read off the corresponding median value.

Accuracy is always a problem with graphical scales but we can see that the corresponding median value is about 18 or £18 000. You might like to draw a more accurate graph and get nearer to the more precise answer of £17 875.

The median – by calculation

The formula is given by:

$$\text{median} = l + i\left(\frac{n/2 - F}{f}\right)$$

where
- l is the lower boundary of the median group
- i is the width of the median group
- F is the cumulative frequency up to the median group

and
- f is the frequency in the median group

By substitution, we can find the median value of £17 875:

$$\text{median} = 15 + 5\frac{(187.5 - 130)}{100} = 15 + 5 \times \frac{57.5}{100} = 17.875$$

The mode

Once we are working with tabulated continuous data, we need to think about the highest point of the distribution rather than the most frequent value. In figure 5.3 we have reproduced the histogram drawn for this data in Chapter 4 (see figure 4.14). You may recall that if the table has unequal interval widths, you need to scale the heights of the blocks (with a histogram you plot frequency in proportion to area).

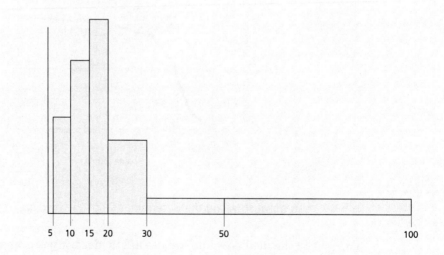

FIGURE 5.3   A histogram drawn for continuous data

It can be seen that the mode will lie in the income range £15 000 to £20 000, and this is referred to as the **modal group** or modal class. To obtain a point estimate of the mode, we first identify the tallest block and then join corner points as shown (see figure 5.4). At the points where these lines cross you can read off the median value (sometimes referred to as the point of maximum density). You might like to draw a more accurate histogram and get nearer to the more precise answer of £16 250.

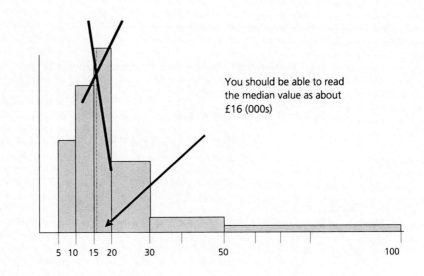

You should be able to read the median value as about £16 (000s)

FIGURE 5.4   Using the histogram to determine the mode

A formula for the calculation of the mean is given in Jon Curwin and Roger Slater, *Quantitative Methods for Business Decisions* (Thomson Learning, 2002).

5.4 │ CONCLUSIONS

The mean, median and mode are not the only measures of location but they are the most common. The mean is the most commonly used measure and is referred to as the average. The mean, median and mode all give different information and in that sense are complementary. The mean can be thought of as giving the 'centre of gravity', the median divides the distribution in two (50 per cent are below the value and 50 per cent above) and the mode gives the 'most likely' value (see figure 5.5).

The relative positions of the mean, median and mode also tell us something about the distribution of the data. Typically, income or wealth data will have a positive skew, with a few individuals on very high income or very wealthy. The

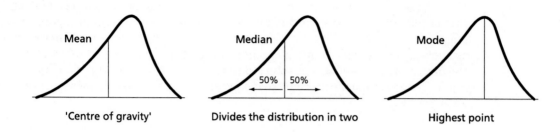

FIGURE 5.5   Interpreting the mean, median and mode

few high values will tend to pull the mean upwards. In our example the mean was £22 400, the median £17 875 and the point estimate of the mode £16 250, suggesting a positively skewed distribution. This skew can also be seen in the histogram drawn in figure 5.4.

The relationship between the mean, median and mode is shown in figure 5.6.

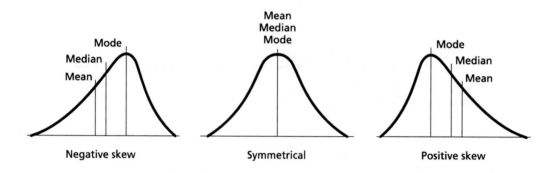

FIGURE 5.6   The relationship between the mean, median and mode and the shape of the distribution

When you have worked through and understood the chapter you will be in a position to

- calculate statistics like the mean, median and mode for different types of data
- comment on the usefulness of these statistics for particular problems
- appreciate the benefits of using a computer package, like Excel, for the computations
- provide useful summary measures of 'average' that could be used as part of a presentation or analysis

1 You have been given the following estimates for the cost of repairing a computer:

£84.00, £72.00, £100.00, £100.00, £68.00

Determine the mean, median and mode.

2 A sample of eight students gave the following times in response to a question on travel time to college:

15 mins, 45 mins, 5 mins, 5 mins, 30 mins, 5 mins, 1hr 15 mins, 55 mins

Determine the mean, median and mode.

3 You have been given the responses of 17 students to a question on how much they have spent on textbooks in the last month:

0	0	£19.99	£32.98	£19.99	£19.99
£48.20	£32.98	0	£19.99	0	£32.98
£19.99	0	£24.50	0	£32.98	

Determine the mean, median and mode. Comment on your results.

4 An 85-page document has just been proof read and the number of errors found on each page recorded. Given the following table, determine the mean, median and mode.

Number of errors	Number of pages
0	32
1	28
2	18
3	4
4	3

5 You have been asked to investigate the issue of limited parking at a block of flats and have been given the following table detailing the number of vehicles registered for each flat. Determine the mean, median and mode.

Number of vehicles	Number of flats
0	8
1	29
2	12
3	0
4	1

6 A small survey produced the following results on the purchase of credit for 'pay as you go' mobile phones:

Credit purchased	Number of respondents
£10	60
£20	35
£50	8
£100	2

Determine the mean, median and mode.

7 The staff working in a restaurant were asked about the tips they received during a typical week. Their (numerical) replies are summarized in the following table:

Weekly tips	Number of staff
under £1	6
£1 but under £5	6
£5 but under £10	4
£10 but under £20	2
£20 and over	2

Determine the mean, median and mode.

8 The errors identified in a number of transactions have been recorded in the following table. Those that have meant a loss are recorded as a negative and those that give a gain as a positive.

Error	Number of transactions
−£100 but less than −£50	2
−£50 but less than −£10	7
−£10 but less than 0	16
0 but less than £10	14
£10 but less than £50	3
£50 but less than £100	1

Determine the mean, median and mode.

✓

5.6 ANNOTATED ANSWERS

1 To find the mean we add the estimates given and divide by 5:

$$\bar{x} = \frac{84.00 + 72.00 + 100.00 + 100.00 + 68.00}{5} = \frac{424}{5} = £84.80$$

The median is the middle value of the ordered list (so we first put the numbers into order):

£68.00, £72.00, £84.00, £100.00, £100.00

In this case the median is £84.00.

The mode is the most frequent value, in this case £100.00.

These summary statistics are all different and all give different information about the data.

2 We always need to ensure that we are working with the same units, so 1 hr 15 mins would be transformed to 75 mins.

$$\bar{x} = \frac{15 + 45 + 5 + 5 + 30 + 5 + 75 + 55}{8}$$

$$= \frac{235}{8} = 29.375$$

The median is the middle value of the ordered list:

5   5   5   15   30   45   55   75

The median position is found using $(n + 1)/2 = 9/2 = 4.5$. We take the half-way point between 15 and 30 for the median. The median is therefore 22.5.

The mode (most frequent) is 5 mins.

The statistics indicate very clearly a positive skew in the data. We also need to be careful that we are not using such summary statistics to describe two distinct groups rather than one. In this case, one group of students may live in a nearby hall of residence (a 5-minute walk) and others may live at home.

3. This data is dominated by the six respondents that gave a zero answer. In terms of questionnaire design, it would have been better to ask, first, whether they had bought any textbooks in the last month (a filter question) and then ask about amounts. We also need to be careful about the timing of such surveys, because the answers could differ considerably if the 'last month' was August rather than September.

We also need to ensure that the questions serve the purpose of the research. It can be seen that five students spent £19.99 – was this one the recommended textbook? In this example, it might have been useful also to ask about the number of textbooks bought.

Try using Excel to get your answers:

4 To determine the mean and median, columns have been added to give totals (*fx*) and cumulative frequency.

Number of errors (x)	Number of pages (f)	fx	Cumulative frequency
0	32	0	32
1	28	28	60
2	18	36	78
3	4	12	82
4	3	12	85
	85	88	

The mean is 1.035 ($\bar{x} = 88/85$).

The median corresponds to the 43rd ordered observation $(85 + 1)/2$. The median lies in the group with 60 or less observations and is therefore 1.

The mode is 0.

5

Number of vehicles (x)	Number of flats (f)	fx	Cumulative frequency
0	8	0	8
1	29	29	37
2	12	24	49
3	0	0	49
4	1	4	50
	50	57	

The mean is 1.14 ($\bar{x}$ = 57/50)

The median corresponds to 25½ᵗʰ ordered observation (50 + 1)/2. The median lies in the group with 37 or less observations and is therefore 1.

The mode is 1.

The measures all indicate that most flats have one vehicle registered. We might be more concerned about the single flat that has four vehicles registered and the number of unregistered vehicles using the limited parking.

6

Credit purchased £s (x)	Number of respondents (f)	fx	Cumulative frequency
10	60	600	60
20	35	700	95
50	8	400	103
100	2	200	105
	105	1900	

The mean is 1900/105 = £18.10 (rounded to 2 decimal places).

The median corresponds to 53ʳᵈ ordered observation (105 + 1)/2. The median lies in the group with 95 or less observations and is therefore £20.00.

The mode is £10.00.

7 To determine the mean and median, columns have been added to give totals (fx) and cumulative frequency.

Weekly tips	Number of staff (f)	Mid-point (x)	fx	Cumulative frequency
under £1	6	0.50	3.00	6
£1 but under £5	6	3.00	18.00	12
£5 but under £10	4	7.50	30.00	16
£10 but under £20	2	15.00	30.00	18
£20 and over	2	25.00*	50.00	20
	20		131.00	

* assumed mid-point value

$$\bar{x} = \frac{131}{20} = £6.55$$

(rounded to the nearest pence)

$$\text{median} = 1 + 4\frac{(10-6)}{6} = £3.67$$

mode (to be determined from the histogram) = £0.57

The results indicate a very positively skewed distribution. We might also be concerned about the accuracy of responses to a question on a sensitive topic like the amount received as tips.

It should be noted that the mid-point for the final group (£20 and over) has been assumed and other assumed values would lead to different answers.

8

Error (x)	Number of transactions (f)	Mid-point (x)	fx	Cumulative frequency
−£100 but less than −£50	2	−75	−150	2
£50 but less than −£10	7	−30	−210	9
£10 but less than 0	16	−5	−80	25
0 but less than £10	14	5	70	39
£10 but less than £50	3	30	90	42
£50 but less than £100	1	75	75	43
	43		−205	

$$\bar{x} = \frac{-205}{43} = -£4.77$$

$$\text{median} = -10 + 10\frac{(21.5 - 9)}{16} = -10 + 7.8125 = -2.1875$$

$$= -£2.19 \quad \text{(to nearest pence)}$$

mode (to be determined from the construction of the histogram) = −£1.23.

These measures of location again give some measure of the skew within the data. In this case, we would naturally talk about the average loss of £4.77. One of the benefits of using the mean is that we can scale up our findings. If, for example, we asked to estimate the losses on one hundred transactions, we could multiply up (by 100) to give the estimated loss of £477.

WEB REFERENCE 👁

www.thomsonlearning.co.uk/businessandmanagement/curwin3

See the companion web site for further questions and annotated answers. There you will also find a PowerPoint presentation which you can use to help understand this area, or later, for revision. The site also contains links to other sites of interest.

Chapter 6

MEASURES OF DIFFERENCE

In Chapter 5 we considered some of the measures of average. We recognized that the mean was particularly important but also that the median and mode could provide useful additional information. Taken together they provide a useful summary of what is typical and say something about the skew of the distribution.

In summarizing data, we also want to say something about the differences. In most research we are not just interested in the average but how people or items differ from the average. To know the largest and smallest values can sometimes be sufficient but generally we will want to know what sort of differences to expect.

OBJECTIVES

After reading this chapter and working through the exercises you should be able to:

- describe the differences seen in data using appropriate summary measures
- calculate the range, quartile deviation and standard deviation for various types of data
- use a spreadsheet to determine standard deviation and other statistics
- know when to use such measures of difference

111

6.1 | MEASURES OF DIFFERENCE FOR A SIMPLE LIST OF DATA

We can again consider the salaries of five students who have recently been offered a one-year placement (see section 5.1):

£12 000   £9600   £12 000   £10 500   £13 100

The highest and lowest values

It is useful to know the largest and smallest values. In this case the highest value is £13 100 and the lowest value £9600. These numbers immediately give you some idea about the differences in the data and the extremes you are likely to find. They also give you a quick error check. If a mistake is made, particularly in losing or gaining digits this will often show as an extreme value.

The **range** is just the difference between the largest and smallest values. In this case the range is equal to:

range = £13 100 – £9600 = £3500

The standard deviation

The **standard deviation** is a measure of the average difference from the mean. The mean for this data is:

$$\bar{x} = \frac{57\,200}{5} = £11\,440$$

The differences about the mean $(x - \bar{x})$ are shown in figure 6.1 below:

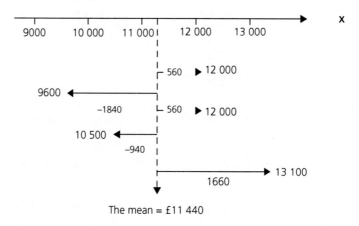

The mean = £11 440

FIGURE 6.1   Deviations about the mean

The differences below the mean are negative and those above are positive. If you were to add these differences, you would find that the sum would come to zero

$$560 - 1840 + 560 - 940 + 1660 = 0$$

To merely average the differences about the mean does not give us a measure of spread (they cancel out). This cancelling out is sometimes referred to as the 'sign problem'. To overcome this sign problem, these differences are squared in the calculation of standard deviation.

The standard deviation is calculated as follows:

1 determine the mean, $\bar{x}$
2 calculate the differences about the mean, $(x - \bar{x})$
3 square these differences to make them all positive, $(x - \bar{x})^2$
4 sum these squared differences, $\Sigma (x - \bar{x})^2$
5 average the squared differences to find **variance**,

$$\frac{\sum (x - \bar{x})^2}{n}$$

6 square root variance to find standard deviation,

$$\sqrt{\frac{\sum (x - \bar{x})^2}{n}}$$

The calculations are shown below in table 6.1:

x	$(x - mean)$	$(x - mean)^2$
12 000	560	313 600
9600	−1840	3 385 600
12 000	560	313 600
10 500	−940	883 600
13 100	1660	2 755 600
		7 652 000

TABLE 6.1   Calculations of standard deviation

The variance

$$= \frac{7\,652\,000}{5} = 1\,530\,400$$

The standard deviation

$$= \sqrt{\frac{7\,652\,000}{5}} = £1237.09$$

If you are given variance you only need to square root to find the standard deviation.

The standard deviation is useful for looking at the variation within a sample of data and for making comparisons between different data sets. The standard deviation uses all the data whereas the range only uses the two extreme values.

If you were given a second data set:

£11 500   £9600   £11 500   £11 500   £13 100

The range would remain the same (£13 100 – £9600) but the standard deviation would be reduced to £1109.24, reflecting the fact that there is less spread in the data. Try to calculate this standard deviation.

The determination of the standard deviation using Excel is shown in figure 6.2.

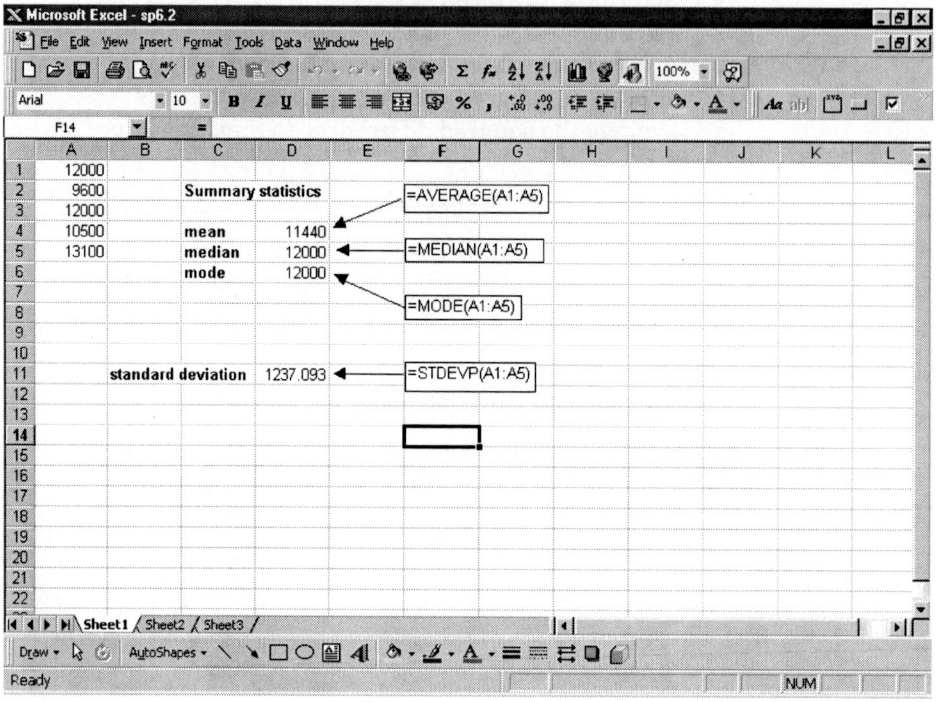

FIGURE 6.2   A spreadsheet showing the standard deviation

We have chosen to calculate the standard deviation using a divisor of n. There are good reasons (not covered by this book) for using a divisor of $n - 1$. As the sample gets larger the difference between using n and $n - 1$ becomes smaller anyway, and we don't want to drift into a technical argument. The point here however, is that the answer you will get from Excel will depend on whether the function you choose for the calculation of the standard deviation uses n or $n - 1$ (as you can see in figure 6.2 we have used STDEVP).

In section 5.1 we considered the determination of the median (the value in the middle location). If this list of numbers were longer we could usefully find other positional statistics, such as the **lower quartile** (the value ¼ of the way along) and the **upper quartile** (the value ¾ of the way along).

Suppose we are now given the salaries of 15 placement students as shown in table 6.2:

£12 000	£9600	£12 000	£10 500	£13 100
£10 400	£12 200	£9200	£6900	£12 250
£12 800	£12 000	£9800	£11 800	£12 100

TABLE 6.2   Salaries of 15 placement students

The largest value is unchanged at £13 100 but the lowest value is now £6900. This lowest value of £6900 may raise some interesting questions. It might be that the salary is lower for accepted reasons, like the placement offers the benefits of living and working abroad but cannot pay a more commercial pay rate. It might be that the placement is only for six months, which would raise the further question of whether this time was sufficient to meet the course requirements for placement. What is important is that data becomes informative.

Excel was used to find the mean = £11 110.00, the median = £12 000, the mode = £12 000 and the standard deviation = £1620.31.

To find the upper and lower quartiles we use $\frac{1}{4}(n + 1)$ and $\frac{3}{4}(n + 1)$. For lists of numbers we still use $(n + 1)$ to work out location. Given that $n = 15$, the lower quartile will correspond to the 4th value ($\frac{1}{4}$ of 16) and the upper quartile to the 12th value ($\frac{3}{4}$ of 16).

If the list is now ordered as below in table 6.3:

£6900	£9200	£9600	£9800	£10 400
£10 500	£11 800	£12 000	£12 000	£12 000
£12 100	£12 200	£12 250	£12 800	£13 100

TABLE 6.3   Ordering the list of salaries of 15 placement students

The lower quartile is £9800 (the 4th order value) and the upper quartile is £12 200 (the 12th ordered value).

6.2   MEASURES OF DIFFERENCE FOR TABLES OF DISCRETE DATA

Results are often presented in the form of a tabulated discrete data and this form is shown below in table 6.4.

Number of children	Number of families
0	40
1	30
2	20
3	8
4	2

TABLE 6.4   Tabulated discrete data on number of children in each family

We can provide some description of this data: the mean is 1.02, the median is 1 and the mode is 0 (see section 5.2 for calculations). However, we may also be interested in how families differ in size.

The range

The range is still the difference between the largest and smallest observation. In this case since some families have 4 children and other families have 0 children, the range is therefore 4.

The quartiles

The median is only one measure of position (the middle value). In this example, the median is 1. We can also examine cumulative frequency to identify other measures of position (order statistics). We find the value ¼ of the way along (the lower quartile) and the value ¾ of the way along (the upper quartile). Given 100 families (see table 6.5), the lower quartile will be 0 corresponding to the 25th ordered observation and the upper quartile will be 2 corresponding to the 75th ordered observation.

(Note: we could use $\dfrac{n+1}{4}$ rather than $\dfrac{n}{4}$ to be more precise, but given the sample size, the effect is negligible.)

The difference between the upper and lower quartile is known as the *interquartile range* and is equal to $2 - 0 = 2$.

❗ Check that you can find these corresponding values by working down the cumulative frequency column in table 6.5 below. You need to identify the row which contains your order observation of interest.

Number of children	Number of families	Cumulative frequency
0	40	40
1	30	70
2	20	90
3	8	98
4	2	100

TABLE 6.5   Cumulative frequency of number of children in each family

We can also find **percentiles** a position defined by a percentage. In this case, the 95th percentile will also correspond to the 95th family and we could make a statement like 'the largest 5 per cent of families have three or more children'.

The standard deviation

To calculate the standard deviation we need to add a step to allow for frequencies:

1   determine the mean, $\bar{x}$
2   calculate the differences about the mean, $(x - \bar{x})$
3   square these differences to make them all positive, $(x - \bar{x})^2$
4   allow for frequencies, $f(x - \bar{x})^2$
5   sum all the squared differences, $\Sigma f(x - \bar{x})^2$
6   average the sum of squared differences to find variance,

$$\frac{\Sigma f(x - \bar{x})^2}{n}$$

7   square root variance to find standard deviation,

$$\sqrt{\frac{\Sigma f(x - \bar{x})^2}{n}}$$

The mean is 1.02. You could check this. The calculations of standard deviation shown in table 6.6 are:

Number of children (x)	Number of families (f)	fx	(x − x̄)	(x − x̄)²	f(x − x̄)²
0	40	0	−1.02	1.0404	41.6160
1	30	30	−0.02	0.0004	0.0120
2	20	40	0.98	0.9604	19.2080
3	8	24	1.98	3.9204	31.3632
4	2	8	2.98	8.8804	17.7608
	100	102			109.9600

TABLE 6.6   Calculations of standard deviation

In our calculations we usually just produce the standard deviation and only give the variance if asked:

$$= \sqrt{\frac{109.9600}{100}} = 1.0486$$

6.3   MEASURES OF DIFFERENCE FOR TABLES OF CONTINUOUS DATA

Results are often presented in the form of tabulated continuous data (you have seen this table in section 5.3 shown here as table 6.7). You may remember that you constructed a histogram to represent this type of data (see section 4.6).

Income group	Number of people
£5000 but under £10 000	50
£10 000 but under £15 000	80
£15 000 but under £20 000	100
£20 000 but under £30 000	80
£30 000 but under £50 000	40
£50 000 but under £100 000	25

TABLE 6.7   Income summarized using unequal intervals

Typically we don't quote a range if we have lost the exact values for the highest and lowest. All that we know now is that all respondents have an income above £5000 and below £100 000.

The quartiles

To find the quartiles and other positional (order) statistics, we need cumulative frequency as shown in table 6.8. We are now working in terms of thousands to make the numbers more manageable (you need to be aware of the capabilities of your calculator).

Income group (£000s)	Number of people (f)	Cumulative frequency
5 but under 10	50	50
10 but under 15	80	130
15 but under 20	100	230
20 but under 30	80	310
30 but under 50	40	350
50 but under 100	25	375
	375	

TABLE 6.8   Income summarized showing cumulative frequency

We plot cumulative frequency against the upper boundary of the corresponding interval. The graph would show that 50 people had an income of £10 (thousand) or less, that 130 people had an income of £15 (thousand) or less and so on. The graph (ogive) is shown as figure 6.3.

The lower quartile will correspond to the 93.75 (using $\frac{1}{4} n$) on the cumulative frequency scale and the upper quartile to the 281.25 (using $\frac{3}{4} n$).

Again it is difficult to read off values with any accuracy from this kind of graph. You might like to draw a more accurate graph and try to get nearer to the more precise answers of £12 730 for the lower quartile and £26 410 for the upper

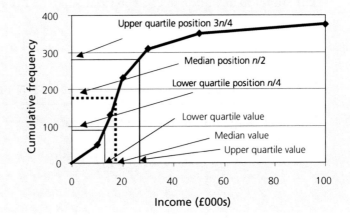

FIGURE 6.3   An ogive showing the determination of quartiles and median

quartile. To calculate these values, you can adapt the formula given for the median (see Jon Curwin and Roger Slater, *Quantitative Methods for Business Decisions*, 5th ed., Thomson Learning, 2002).

The interquartile range is £26 410 – £12 730 = £13 680.

The standard deviation

To calculate the standard deviation we need to use mid-points as our x values. The formula, allowing for frequencies is:

$$\sqrt{\frac{\sum f(x - \bar{x})^2}{n}}$$

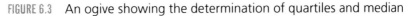

We know the mean to be 22.4 (£000s) or £22 400.

The calculations (working in 000s) are shown below in table 6.9:

Income group (£000s)	Mid-point (x)	No. of people (f)	fx	(x − x̄)	(x − x̄)²	f(x − x̄)²
5 but under 10	7.5	50	375	−14.9	222.01	11 100.5
10 but under 15	12.5	80	1000	−9.9	98.01	7840.8
15 but under 20	17.5	100	1750	−4.9	24.01	2401.0
20 but under 30	22.5	80	1800	0.1	0.01	0.8
30 but under 50	40	40	1600	17.6	309.76	12 390.4
50 but under 100	75	25	1875	52.6	2766.76	69 169.0
		375	8400			102 902.5

TABLE 6.9   Income showing standard deviation

The standard deviation is £16.565 (thousand) or £16 565

$$\sqrt{\frac{102\,902.5}{375}} = 16.565224$$

The standard deviation – an alternative means of calculation

You can manipulate the formula used so far:

$$\sqrt{\frac{\sum f(x - \bar{x})^2}{n}}$$

By expanding the brackets and rearranging we get an alternative format that will make calculations by hand easier (see Jon Curwin and Roger Slater, *Quantitative Methods for Business Decisions*, 5th ed., Thomson Learning, 2002):

$$\sqrt{\left[\frac{\sum fx^2}{n} - \left(\frac{\sum fx}{n}\right)^2\right]}$$

The calculations are shown below in table 6.10:

Income group (£000s)	Mid-point (x)	No. of people (f)	fx	fx²
5 but under 10	7.5	50	375	2812.5
10 but under 15	12.5	80	1000	12 500
15 but under 20	17.5	100	1750	30 625
20 but under 30	22.5	80	1800	40 500
30 but under 50	40	40	1600	64 000
50 but under 100	75	25	1875	140 625
		375	8400	291 062.5

TABLE 6.10   Alternative format for standard deviation

Substitution into the formula gives

$$\sqrt{\left[\frac{291\,062.5}{375} - \left(\frac{8400}{375}\right)^2\right]} = 16.565224$$

The same answer – as expected.

You have now seen two ways of calculating the standard deviation. Which you choose will depend on the circumstances and the preference of your course lecturer.

6.4 | CONCLUSIONS

The standard deviation is the most commonly used measure of spread. It has useful qualities like including all the data, making large differences important (the effect of squaring differences) and important applications in more advanced statistics (see Jon Curwin and Roger Slater, *Quantitative Methods for Business Decisions*, 5th ed., Thomson Learning, 2002).

We have used the standard deviation as a descriptive measure: more variable data gives a larger value. It is important to try to understand and interpret the value calculated for standard deviation. The value will change if we change units, e.g. from euros to dollars. If the sample examined is more mixed, then the standard deviation is, likely to be larger, e.g. if we examine the income of skilled and unskilled workers together, then the standard deviation will be larger than looking at each separately.

In market research or quality management we are interested in discovering differences but then need to consider possible explanation of the difference. If you are asked to work with data, the chances are a range of statistics will be provided or needed. You need to ensure that the statistics are appropriate (you cannot have the average political party that people voted for at the last election or standard deviation for gender classification) and provide a mechanism for a better understanding of the phenomena of interest.

When you have worked through and understood the chapter you will be in a position to:

- calculate a number of measures of difference, such as the standard deviation
- comment on the usefulness of such measures and be able to comment on the diversity within the data
- appreciate the benefits of using a computer package, like Excel, for the computations
- provide useful summary measures of the data that could be used as part of a presentation or analysis

6.5 EXERCISES

1 You have been given the following estimates for the cost of repairing a computer:

£84.00, £72.00, £100.00, £100.00, £68.00

Determine the mean, range and standard deviation.

2 A sample of eight students gave the following times in response to a question on travel time to college:

15 mins, 45 mins, 5 mins, 5 mins, 30 mins, 5 mins, 1hr 15 mins, 55 mins

Determine the mean, range and standard deviation.

3 You have been given the responses of 17 students to a question on how much they have spent on textbooks in the last month:

0	0	£19.99	£32.98	£19.99	£19.99
£48.20	£32.98	0	£19.99	0	£32.98
£19.99	0	£24.50	0	£32.98	

Determine the mean and standard deviation. Comment on your results.

4 An 85-page document has just been proof read and the number of errors found on each page recorded. Given the following table, determine the mean, median, quartiles, interquartile range and standard deviation.

Number of errors	Number of pages
0	32
1	28
2	18
3	4
4	3

5 You have been asked to investigate the issue of limited parking at a block of flats and have been given the following table detailing the number of vehicles registered for each flat. Determine the mean and standard deviation.

Number of vehicles	Number of flats
0	8
1	29
2	12
3	0
4	1

6 A small survey produced the following results on the purchase of credit for 'pay as you go' mobile phones:

Credit purchased	Number of respondents
£10	60
£20	35
£50	8
£100	2

Determine the mean and standard deviation.

7   The staff working in a restaurant were asked about the tips they received during a typical week. Their (numerical) replies are summarized in the following table:

Weekly tips	Number of staff
under £1	6
£1 but under £5	6
£5 but under £10	4
£10 but under £20	2
£20 and over	2

Determine the mean, median, quartiles, interquartile range and standard deviation.

8   The errors identified in a number of transactions have been recorded in the following table. Those that have meant a loss are recorded as a negative and those that give a gain as a positive.

Error	Number of transactions
–£100 but less that –£50	2
–£50 but less that –£10	7
–£10 but less that 0	16
0 but less than £10	14
£10 but less than £50	3
£50 but less than £100	1

Determine the mean, median, quartiles, interquartile range and standard deviation.

6.6 ANNOTATED
ANSWERS

1   The calculations are shown in the following table. The mean is £84.80 (see question 5.1).

x	(x – mean)	(x – mean)^2
84.00	–0.80	0.64
72.00	–12.80	163.84
100.00	15.20	231.04
100.00	15.20	231.04
68.00	–16.80	282.24
424.00		908.80

$$\bar{x} = \frac{424}{5} = £84.80$$

If the values are ordered we get:

68.00   72.00   84.00   100.00   100.00

The range = 100.00 – 68.00 = £32.00

The standard deviation:

$$= \sqrt{\frac{908.80}{5}} = £13.48$$

(rounded to two decimal places)

2   The calculations are shown in the following table. The mean is 29.375 minutes (see question 5.2).

x	(x – mean)	(x – mean)^2
15	–14.375	206.6406
45	15.625	244.1406
5	–24.375	594.1406
5	–24.375	594.1406
30	0.625	0.3906
5	–24.375	594.1406
75	45.625	2081.6406
55	25.625	656.6406
235		4971.8748

$$\overline{x} = \frac{235}{8} = 29.375 \text{ mins}$$

If the values are ordered we get:

$$5 \quad 5 \quad 5 \quad 15 \quad 30 \quad 45 \quad 55 \quad 75$$

The range: $75 - 5 = 70$ mins

The standard deviation:

$$= \sqrt{\frac{4971.8748}{8}} = 24.93 \text{ mins}$$

(rounded to two decimal places)

3   Try using Excel to get your answers:

The six respondents that gave a zero answer have an important influence on the answer and we need to decide whether to include all respondents (as here) or just those that purchased textbooks. This brings us back to a discussion of what do we want the mean and standard deviation to tell us!

4   To determine the mean and standard deviation we need to allow for frequencies.

Number of errors (x)	Number of pages (f)	fx	$(x-\bar{x})$	$(x-\bar{x})^2$	$f(x-\bar{x})^2$
0	32	0	−1.035	1.071225	34.2792
1	28	28	−0.035	0.001225	0.0343
2	18	36	0.965	0.931225	16.7621
3	4	12	1.965	3.861225	15.4449
4	3	12	2.965	8.791225	26.3737
	85	88			92.8983

The mean is:

$$\bar{x} = \frac{88}{85} = 1.035$$

(rounded to 3 decimal places)

Standard deviation:

$$= \sqrt{\frac{92.8983}{85}} = 1.0454$$

(rounded to 4 decimal places)

To find the median and quartiles we need to examine cumulative frequency:

Number of errors (x)	Number of pages (f)	Cumulative frequency
0	32	32
1	28	60
2	18	78
3	4	82
4	3	85
	85	

The median corresponds to the 43rd ordered observation (85 + 1)/2. The median lies in the group with 60 or less observations and is therefore 1.

The lower quartile (corresponds to the ¼ × 85 = 21.25 ordered value) is 0

The upper quartile (corresponds to the ¾ × 85 = 63.75 ordered value) is 2

The interquartile range is therefore 2 − 0 = 2

5 Given the following summary statistics (you could construct a spreadsheet to get these), we can calculate the mean and standard deviation as shown below:

$$n = 50$$

$$\sum fx = 57$$

and

$$\sum f(x - \overline{x})^2 = 28.0200$$

The mean:

$$\overline{x} = \frac{57}{50} = 1.14$$

and the standard deviation:

$$= \sqrt{\frac{28.0200}{50}} = 0.7486$$

6 Given the following summary statistics (you could construct a spreadsheet to get these), we can calculate the mean and standard deviation as shown below:

$$n = 105$$

$$\sum fx = 1900$$

and (assuming the rounded mean of £18.10):

$$\sum f(x - \overline{x})^2 = 25\,619.05$$

The mean:

$$\overline{x} = \frac{1900}{105} = 18.10$$

and the standard deviation:

$$= \sqrt{\frac{28.0200}{50}} = 0.7486$$

(rounded to 2 decimal places)

7 Summary statistics can be determined using a spreadsheet or calculator (a mid-point of 25 has been assumed for the last group):

$$n = 20$$

$$\sum fx = 131$$

$$\sum f(x - \overline{x})^2 = 1122.45$$

The mean (see question 5.5):

$$\bar{x} = \frac{131}{20} = £6.55$$

$$\bar{x}\sqrt{\frac{1122.45}{20}} = £7.4915$$

and the standard deviation:

$$= \sqrt{\frac{1122.45}{20}} = £7.4915$$

The quartiles can be estimated from the construction of an ogive (see answer to question 5 in section 5.5) or adapting the formula for the median.

The lower quartile = £0.83

The upper quartile = £8.73

The inter-quartile range = £8.73 – £0.83 = £7.90

8 Summary statistics can be determined using a spreadsheet or calculator:

$$n = 43$$
$$\sum fx = -205$$
$$\sum f(x - \bar{x})^2 = 25\ 647.67$$

The mean:

$$\bar{x} = \frac{-205}{43} = -£4.77$$

and the standard deviation:

$$= \sqrt{\frac{25\ 647.67}{43}} = 24.42$$

(to 2 decimal places)

Using the alternative formula given:

$$\text{standard deviation} = \sqrt{\left[\frac{\sum fx^2}{n} - \left(\frac{\sum fx}{n}\right)^2\right]}$$

Using:

$$\sum fx = -205$$

and

$$\sum fx^2 = 26\,625$$

then:

$$\sqrt{\left[\frac{\sum 26\,625}{43} - \left(\frac{-205}{43}\right)^2\right]} = 24.42$$

which gives the same answer.

The quartiles can be estimated from the construction of an ogive (see answer to question 6 in section 5.5) or adapting the formula for the median.

The lower quartile = –£8.91

The upper quartile = £5.18

The inter-quartile range = £5.18 – (–£8.91) = £14.09

It is unlikely that your course will require the calculations for the upper and lower quartiles but they are included here for completeness:

The lower quartile:

$$-£10 + 10\left(\frac{10.75 - 9}{16}\right) = -£8.91$$

The upper quartile:

$$£0 + 10\left(\frac{32.25 - 25}{14}\right) = £5.18$$

WEB REFERENCE ⊚

www.thomsonlearning.co.uk/businessandmanagement/curwin3

See the companion web site for further questions and annotated answers. There you will also find a PowerPoint presentation which you can use to help understand this area, or later, for revision. The site also contains links to other sites of interest.

Chapter 7

MAKING COMPARISONS USING INDEX NUMBERS

When we look at numbers we often need to make comparisons between them. This might be from one year to another, from one region or country to another, or even from one supplier to another. Here we are going to look at a method of comparison which uses **index numbers**.

OBJECTIVES

After reading this chapter and doing the exercises you should be able to:

- describe the concept of an index number
- explain why they are calculated
- calculate simple index numbers
- time-align index numbers
- calculate Laspeyres indices
- calculate Paasche indices
- describe at least one published index
- find published data and series

7.1 | WHAT IS AN INDEX NUMBER?

Data can be very awkward to compare. Sometimes there have been big changes from one date to another, sometimes we want to compare the changes in different currencies and sometimes we even want to combine data which relate to totally different ways of measuring, such as kilograms and litres. Also, some numbers that we might want to compare are very large, for example, GDP figures, and a simple comparison is then difficult to make.

Whether you are looking at a simple set of figures which all relate to, say, the price of a single item over time, or numbers which try to combine the prices of all of the purchases someone makes, an index number brings everything back to a comparison with the number 100. This is known as the *base*. If this is the case, then comparison becomes easy! Take the following example in table 7.1:

Year	Index of cinema ticket prices	Index of football match ticket prices
1	100	100
2	105	110
3	110	115
4	110	135
5	112	140

TABLE 7.1   Tickets showing the base index number

Over the five-year period, cinema tickets have increased from 100 to 112 – a rise of 12 per cent, whilst football match tickets have gone up by 40 per cent from 100 to 140. The comparison can be made for any particular year, for instance, in Year 3, the respective increases are 10 per cent and 15 per cent from the base year. You should notice that in each case we compare back to Year 1 – the **base year**. Comparison to the base year gives percentage increases. Otherwise we need to talk about the increase in percentage points. We can say that cinema tickets went up by 5 points from Year 2 to Year 3. We could also say that football match tickets went up by 5 points in the same period but this does not give the percentage increase. Simple subtraction only gives the points increase

🛈 See notes in section 2.10 on percentage changes.

7.2 | COMPARISONS WITH A SINGLE SET OF FIGURES

The arithmetic necessary to make these comparisons is simple, but the more difficult task is to choose the **base year**. A base year should be typical, whatever that is, with no significant events. Such events are often not obvious for several years after they take place. On a first-year course you are usually told which year to use as a base year.

An example will help to explain. Given the price of bottled beer over a five-year period, as in table 7.2 below, we can construct an index series to represent the price changes.

Year	Price per bottle
1	1.50
2	1.62
3	1.69
4	1.79
5	2.05

TABLE 7.2   Price of beer over a 5-year period

Let's assume that we are going to use Year 1 as the base year, so we need to give it the value of 100. From then on, we want the percentage increase in the price of a bottle of beer since Year 1. We find this for Year 2 by putting the Year 2 price over the Year 1 price and multiplying the result by 100:

$$\frac{1.62}{1.50} \times 100 = 1.08 \times 100 = 108$$

In other words, there has been an 8 per cent increase in price from Year 1 to Year 2.

For Year 3 we have the following calculation:

$$\frac{1.69}{1.50} \times 100 = 1.126667 \times 100 = 112.7 \text{ (rounded)}$$

Note that we still use the Year 1 price on the bottom line. Finishing off the calculations, we get the index series shown below in table 7.3:

Year	Price per bottle	Index series
1	1.50	100
2	1.62	108
3	1.69	112.7
4	1.79	119.3
5	2.05	136.7

TABLE 7.3   Index series of price change in beer over a 5-year period

This shows a 36.7 per cent increase in the price of bottled beer over the five-year period. You might want to graph these figures, as in figure 7.1, to give a better understanding.

An annotated screen shot from Excel (figure 7.2) shows how you might calculate these figures on a spreadsheet.

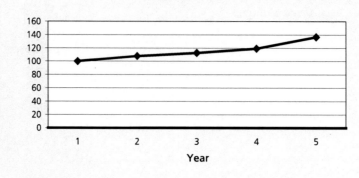

FIGURE 7.1   Graph of bottled beer index

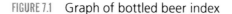
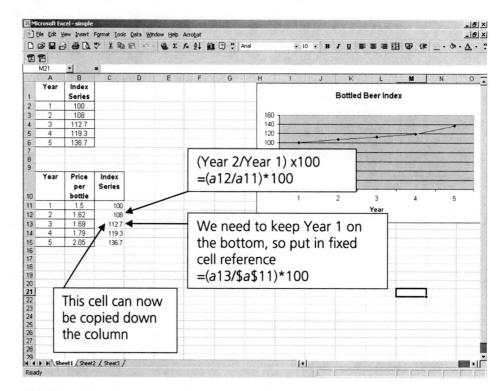

FIGURE 7.2   Using Excel to calculate simple index series

The formula

$$P_n/P_1 \times 100$$

is a reminder of the calculations where P_1 is the base year price and P_n (price now), the price in subsequent years.

7.3 | TIME-ALIGNING SERIES

If all index numbers always started in the same base year, comparisons from one series to another would be completely straightforward. Unfortunately, this is not the case in practice. Although the majority of published series are calculated and produced by the same body, the Office for National Statistics in the UK, they have varying **base years**. For simplicity in making comparisons, we need some method of making all of the base years the same for the series we want to compare.

The first task is to choose one of the years or time periods as the base year that we are going to use. Having done this, we can then convert the other series to the same base year. This is possible since an index number measures the percentage change from its base year, so a change from one year to another must remain proportional when we change the base year. Let's look at the example shown below as table 7.4.

Year	Index of wages	Index of prices
1	100	120
2	104	123
3	110	127
4	117	134
5	121	138
6	128	140
7	135	145

TABLE 7.4 Price and wage variations over a 7-year period

We will take Year 1 as the base year, the figures for wages can stay the same. To convert the figures for prices, we need to make the Year 1 figure into 100. (Remember that these are proportional figures.) To do this we can divide the figure by 120 (the value in our new base year) and multiply the result by 100.

For Year 1:

$$\frac{120}{120} \times 100 = 1 \times 100 = 100$$

For Year 2 we do the same, still using 120 on the bottom line:

$$\frac{123}{120} \times 100 = 1.025 \times 100 = 102.5$$

We can carry on until we have done all of the data. This is known as **rebasing** the index series. The result can be graphed as in figure 7.3 to make the comparison even easier.

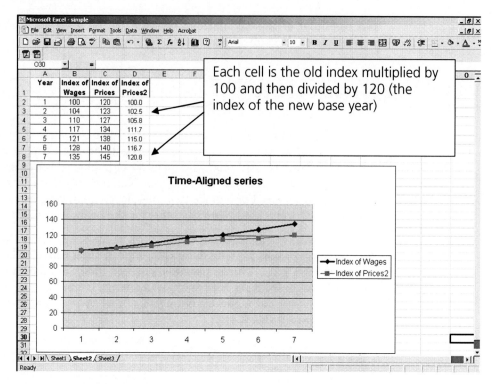

FIGURE 7.3   Graphing the two series

Looking at the figures, or the graph, it is now quite easy to see that over this time period, wages have increased more than prices. This is usually interpreted as in increase in the standard of living.

We can extend this idea to join two (or more) series together. Obviously the series must be measuring the same thing, but what often happens is that a series is rebased, and so starts from 100 again. Look at the example in table 7.5:

Year	Old index	New index
1	140	
2	169	
3	187	
4	197	100
5		110
6		124
7		134
8		140
9		145

TABLE 7.5   Rebasing a series

The key factor in joining the two series together is having an old and a new value for a single year; here Year 4. Remember that index numbers are ratios or percentage changes from the base year, so to get from one series to the other we need to use multiplication and division. To make 100 into 197 we need to multiply by 1.97. Since all the new index figures are relative to this number, we can just multiply all of the data in the old index by 1.97. The result is shown below in table 7.6:

Year	Old index	New index	Combined index
1	140		140
2	169		169
3	187		187
4	197	100	197
5		110	216.7
6		124	244.28
7		134	263.98
8		140	275.8
9		145	285.65

TABLE 7.6   Combining the index by converting the new series back to the old series

We could do it the other way around and convert the old series to the new one. To make 197 into 100 you divide by 1.97, and doing this to the rest of the old series is shown in table 7.7:

Year	Old index	New index	Combined index
1	140		71.07
2	169		85.79
3	187		94.92
4	197	100	100
5		110	110
6		124	124
7		134	34
8		140	140
9		145	145

TABLE 7.7   Combining the index by converting the old series to the new series

7.4 COMPARISONS WITH GROUPS OF FIGURES

Looking at just one lot of figures may hold some interest, but index numbers are most important because they are able to amalgamate the variations in a whole group of items into a single index series. Perhaps the most spectacular version of this is the *retail prices index* where several hundred prices are combined together – see section 7.7. Calculations for groups of figures are basically easy, but very boring if you have to do a lot of them by hand, since you keep doing the same thing over and over again. Spreadsheets are the ideal way to do the calculations, since you can put the formula in once and then just copy as many times as you need.

7.5 BASE WEIGHTING

When we are dealing with a group of items, say everything a company buys from a particular supplier, or the yield of various crops on a farm, we cannot just add them all together and work with the totals.

 Can you see why not?

Just imagine for a moment that you have two part-time jobs. In the first you work for two hours and get €100 per hour. In the second you work for ten hours and get €20 per hour. Do you think it fair if someone said that you were earning €60 ((100 + 20)/2) per hour? Obviously not. If that were the case, you would be picking up €720, rather than the €400 you actually get paid! Simple averaging didn't work because you worked for different amounts of time at each job. What we would need to do is weight the earnings from each job by the amount that you worked. (This would give an average of €33.33 per hour: 400/12.) Index numbers are really a type of averaging.

The most obvious way to combine figures over time is to use a **base-weighted** system, since these are the figures we are most likely to have. Let's work through an example (set out below as table 7.8) to get the idea, then the main calculations can be done on a spreadsheet.

Item	Price (Yr1)	Quantity (Yr1)	Price (Yr2)	Price (Yr3)
Chips	2	7	2.1	2.2
Beer	4	20	4.6	4.9
Cigarettes	8	4	9.2	10.5

TABLE 7.8   Student purchases

If we just look at the figures for the first year, and also decide to use this as our base year, we can easily calculate the amount spent on each item by multiplication. You should get a total of 126.

If all of the information that we have is in table 7.8, then we can continue to use the quantities when we work out the spending in Year 2. This will give us the results shown in table 7.9:

Item	Quantity (Yr1)	Price (Yr2)	Amount
Chips	7	2.1	14.7
Beer	20	4.6	92.0
Cigarettes	4	9.2	36.8
Total spend			143.5

TABLE 7.9   The cost of the base year goods in year 2

The index number we need is found by putting the second figure over the first, and multiplying by 100.

$$\frac{143.5}{126} \times 100 = 1.1388888 \times 100 = 113.89 \text{ (rounded)}$$

This shows a 13.89 per cent increase in prices between the two years.

❗ Why just prices? It must be for prices because we used the same quantity figures in both calculations, so they didn't change at all.

Going through the same process for Year 3 gives a total spend figure of 155.4 and doing the division gives an index of 123.3.

❗ Make sure you can get these answers too.

Even though it is very simple data, we have constructed a spreadsheet to illustrate how you could automate many of the calculations and this is shown in figure 7.4. Although our spreadsheet only has three items in the group, you can see that, once the data has been entered, it makes virtually no difference to how long it then takes to calculate the index series since most cells contain formulae which are just copied down a column or across a row.

What we have just calculated is also called a **Laspeyres index** series.

The formula for a Laspeyres price index would be:

$$Laspeyres\ Price\ Index\ Year\ 2 = \frac{\sum (P_2 Q_1)}{\sum (P_1 Q_1)} \times 100$$

where Year 1 is the base year and P_1 and Q_1 are the prices and quantities respectively in the base year. Similarly, P_2 represents the prices in Year 2. For subsequent years you can use P_3, P_4, etc. This type of index compares the amount spent in Year 1 with the amount spent on the same amounts of the same goods in subsequent years.

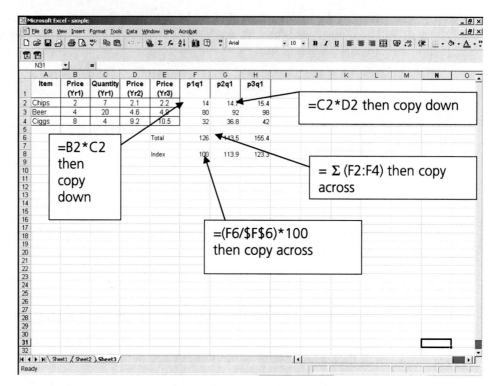

FIGURE 7.4   Spreadsheet to calculate a base-weighted index number

7.6 CURRENT WEIGHTING

Calculating a base weighted index series (shown above in figure 7.4) is usually quicker and normally cheaper than any other sort of index, since we need less data. However, there is a fundamental assumption contained in the calculations. This is that the pattern of expenditure (that is the relative importance of each item in the group) remains the same over time. This might be true in the short term, but is very unlikely over several years.

If we extend the previous example, but now add that our student gave up smoking at the end of Year 1, but as a consequence, began to eat more chips and drink more beer, then we might have the data shown in table 7.10:

Item	Price (Yr1)	Quantity (Yr1)	Price (Yr2)	Quantity (Yr2)	Price (Yr3)	Quantity (Yr3)
Chips	2	7	2.1	10	2.2	10
Beer	4	20	4.6	25	4.9	30
Cigarettes	8	4	9.2	0	10.5	0

TABLE 7.10   Student purchases and prices for Years 1 and 2

Now we have enough information to calculate the amount actually spent per week in each of the three years. We find this by multiplying the actual amount bought (quantity for the week) by the actual price that year. This will give us the following table, as shown in table 7.11.

Item	P1Q1	P2Q2	P3Q3
Chips	14	21	22
Beer	80	115	147
Cigarettes	32	0	0
Totals	126	136	169

TABLE 7.11   Actual spending

Now we need something to compare these figures to, but we only want the prices to vary, so we have to keep the quantities fixed for each year we want to compare. So what we will have to do is work out what would have been spent if we use the base year prices and the current year's quantities. Sounds a bit complicated, but is okay if you take it one step at a time as below in table 7.12:

Item	P1Q2	P1Q3
Chips	20	20
Beer	100	120
Cigarettes	0	0
Totals	120	140

TABLE 7.12   Spending using base year prices

Now all we need to do is to compare the actual spend for a particular year with the potential spend (i.e. using base year prices). For Year 2 we have:

$$\frac{136}{120} \times 100 = 1.1333 \times 100 = 113.33$$

and for Year 3 the calculation is:

$$\frac{169}{140} \times 100 = 1.2071 \times 100 = 120.71$$

This index number is also called a **Paasche index** series.

A formula for a Paasche price index would be:

$$\text{Paasche Price Index Year 2} = \frac{\sum (P_2 Q_2)}{\sum (P_1 Q_2)} \times 100$$

where Year 1 is the base year and P_1 and Q_1 are the prices and quantities respectively in the base year. Similarly, P_2 and Q_2 represent the prices and quantities respectively in Year 2. For subsequent years you can use P_3 and Q_3, P_4 and Q_4, etc.

We have illustrated the calculation of price index numbers, but the formulae can be used, with very slight modification, to calculate quantity indices; see the formula sheet at the back of this text.

7.7 | RETAIL PRICES INDEX

This is probably the most quoted, and misunderstood, index number in the UK. It is a measure of consumer price inflation and relates to the vast majority of the population. It is calculated on a monthly basis and is based on a 'basket of goods' deemed to be typical. Typical is defined from the results of the rolling Expenditure and Food Survey. It is an 'annually chain-linked Laspeyres index' which means that the weights are set for a year and then revised for the following year. Approximately 140 000 prices are checked each month in compiling the index.

Why is it calculated? It is a major factor taken into account by the Treasury and the Bank of England in managing the economy and is viewed as a measure of success by governments and critics alike. It is also linked directly to certain benefits, such as state pensions and is often quoted in wage negotiations.

Further current details can be found on the Office of National Statistics web site at **http://www.statistics.gov.uk** by searching for Retail Prices Index. You can also download raw data from this site. Figure 7.5 shows the official site details for the RPI.

7.8 | PUBLISHED SERIES

There is a tremendous amount of published data (see Chapter 13) and much of it is in index form to make it easier to understand. Often where data is given as actual figures, for example the gross domestic product of a country, it is also given as an index series. For the UK, the first source to check for data is the web site for the government statistical service at **http://www.statistics.gov.uk**

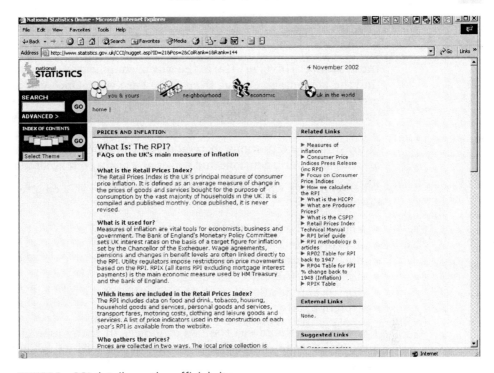

FIGURE 7.5   RPI details on the official site

where you may be able to download data, as we suggested for the RPI. Even if the data is not immediately available to download for free, you can both check on what is available and on details of how it is calculated. If you do not have access to the web, then a visit to a library and a look through the various statistical publications will yield the same information, but perhaps more slowly.

These published series are used in a variety of contexts. They might be used by a company to compare their prices with the general level of inflation. They might be used by a group campaigning on behalf of pensioners to highlight the reducing value of the state pension, or by a trade union in deciding upon the level of a wage claim. Some companies build mathematical and statistical models of their markets, whilst the Treasury builds a model of the economy; both of which will use such published data. It is not only the government who produce index numbers, they may feature in research reports, internal company documents, and marketing analyses. These may be more difficult to obtain unless you work for that particular company.

7.9 | CONCLUSIONS

Index numbers involve little in the way of complicated arithmetic and just try to find the percentage change from some base year. Once you are dealing with a group of items, rather than a single one, this task becomes a little more difficult, but remains fairly easy if you are careful. For more than a few numbers, it is worth using a spreadsheet to do the calculations. Index numbers crop up in many different situations in relation to the economy, valuing investments or assets, calculating pensions and wages or working out acceptable interest rates.

When you have worked through and understood this chapter, you will be in a position to:

- understand published index series
- interpret those series for others
- find percentage changes in published series
- discuss and calculate base-weighted indices
- discuss and calculate current-weighted indices
- explain to others the retail prices index

Typically you would see the use of indices in many management accounting departments and marketing departments, as well as in academic disciplines such as economics.

7.10 EXERCISES

1 Find the simple index numbers for the prices of a special bread using Year 1 as the base year.

Year	Price
1	0.80
2	1.00
3	1.10
4	1.19
5	1.23
6	1.28
7	1.32
8	1.37
9	1.44
10	1.50

2 Find the simple price index for the price of petrol using Year 4 as the base year.

Year	Price per litre
1	1.02
2	1.07
3	1.13
4	1.18
5	1.27
6	1.32
7	1.36
8	1.38
9	1.43

3 Produce a simple quantity index for the amount of illegal booze seized by Customs in a particular region. Use Year 1 as the base year.

Year	Quantities
1	450 000
2	525 000
3	659 000
4	739 500
5	745 900
6	847 800
7	990 000
8	1 025 000

4 You have been asked to compare the increases in wages and prices for a country over a ten-year period. You have been able to collect the following indices, but the wage data has a base year in Year 1, whilst the prices data has its base year as Year 5.

Year	Wages	Prices
1	100	90.5
2	104.3	92.1
3	105.9	95.3
4	108.4	97.5
5	110.5	100
6	114.6	105.3
7	124.6	110.6
8	129.2	113.9
9	133.6	120.4
10	140.3	125.6

Construct a graph to aid understanding and assess if the standard of living has changed during the 10-year period.

5   Some people believe that pay and productivity are linked. You have been asked to compare the two over a lengthy period. The pay index is based in Year 2 and the productivity index at some point in the past. Rebase the productivity index to Year 2 and then graph the two series. Comment on your graph.

Year	Pay	Productivity
1	90	110
2	100	108
3	111	105
4	120	106
5	126	109
6	132	110
7	138	112
8	142	115
9	149	118
10	159	117
11	172	116
12	183	116
13	196	115
14	205	118

6   Britain and the United States both measure domestic output, but in slightly different ways. If we ignore any methodological differences in approach we may compare the output levels. Given the data below, rebase all of the series to Year 5 and produce a graph to contrast the outputs of the two countries. What warning would you offer others looking at your graph?

Year	UK output	UK output	US output
1	102.3		125.4
2	103.5		130.7
3	104.6		140.2
4	107.2		145.3
5	110.4	100	142.1
6		104.2	141.2
7		107.4	145.8
8		106.5	150.5

Continued

Year	UK output	UK output	US output
9		108.2	153.2
10		110.7	148.4
11		110.4	145.3
12		112.6	146.1

7 As an attempt at considering the rate of price increase for basic items, you have been asked to construct an index using the following figures and a base year of Year 1. Calculate your answers and comment on your findings.

Item	Price Yr1	Quantity Yr1	Price Yr2	Price Yr3
Potatoes	0.2	5	0.17	0.15
Coffee	3.4	1	3.8	4.1
Sausages	1.9	3	2.1	2.3

8 Some steel producers have complained that their prices are too low. You have been asked to investigate and have been given the following data on costs and the steel price index:

Costs					
Item	Price Yr1	Quantity Yr1	Price Yr2	Price Yr3	Price Yr4
Coal	30	120	28	26	22
Ore	10	150	11	11	13
Scrap	2	30	2.5	3.1	3.2
Limestone	1	110	1.2	1.3	1.5
Labour	20	40	22	23	24

Year	Steel price index
1	100
2	103
3	105
4	103

Construct a base weighted index of steel producers' costs and draw a comparison with the index of prices. Do they have a case?

9 Five items have been monitored for price over a four-year period and the collected data is as follows:

Item	Price Yr1	Quantity Yr1	Price Yr2	Price Yr3	Price Yr4
A	2.5	50	3	3.6	3.7
B	3.8	40	4.5	5.5	5.5
C	5.3	20	6.4	7.5	7.9
D	6.9	15	7.8	8.6	8.9
E	10.2	8	11.4	12.4	12.7

Calculate an all-items Laspeyres index with Year 1 as the base year.

10 The problem posed in Question 7 has been extended by finding data on quantities purchased in Years 2 and 3. Calculate a current weighted index with Year 1 as the base year. Compare your answer to the one you obtained for Question 7 and comment.

Item	Price Yr1	Quantity Yr1	Price Yr2	Quantity Yr2	Price Yr3	Quantity Yr3
Potatoes	0.2	5	0.17	6	0.15	4
Coffee	3.4	1	3.8	2	4.1	3
Sausages	1.9	3	2.1	2	2.3	2

11 A company has asked you to look at its costs and has provided the following information:

Item	Year 1 Price	Year 1 Quantity	Year 2 Price	Year 2 Quantity	Year 3 Price	Year 3 Quantity
Material 1	2	256	2.4	260	2.4	220
Material 2	5.3	45	5.4	50	5.5	70
Material 3	1.3	400	1.4	500	1.4	600
Labour	8.23	121	8.78	115	9.55	100
Utilities	563.8	3	600.5	3	634.77	3
Local Taxes	886.2	1	1003.67	1	1134.66	1

Use this information to construct a current weighted price index for Years 2 and 3 with Year 1 as the base year.

12 Six items are purchased on a regular basis and their prices and quantities have been recorded for three years. Use the data presented below to find all items price indices for Years 2 and 3 with Year 1 as the base year.

	Year 1		Year 2		Year 3	
Item	Price	Quantity	Price	Quantity	Price	Quantity
A	12	55	13	65	14	65
B	11	120	11.1	120	11.2	130
C	9	150	9.6	140	9.9	150
D	6	220	6.2	200	6.3	190
E	4	250	4.1	260	4.1	280
F	2	900	1.4	1100	1	1200

13 Use the government web site or a library to obtain a list of ten different published index series. Make a note of the base year and the current value of the indices. Suggest where each series would be most useful.

14 Download the RPI data from **www.statistics.gov.uk** and create a graph of the annual inflation from 1949 to the most recent year. How would you interpret this result?

15 Use the data from Question 12 (reproduced below) to find an all-items quantity index for Years 2 and 3 with Year 1 as a base year.

	Year 1		Year 2		Year 3	
Item	Price	Quantity	Price	Quantity	Price	Quantity
A	12	55	13	65	14	65
B	11	120	11.1	120	11.2	130
C	9	150	9.6	140	9.9	150
D	6	220	6.2	200	6.3	190
E	4	250	4.1	260	4.1	280
F	2	900	1.4	1100	1	1200

✓

7.11 ANNOTATED ANSWERS

Note: *Since index number calculations often involve repetitive calculations on columns of figures they are ideally suited to being calculated using a spreadsheet. Most answers here are illustrated using spreadsheets, but the calculations can be done by hand as well.*

1 In the base year the bread costs £0.80 so we want to relate all of the other prices to this figure. For Year 2 we have (£1.00/£0.80) × 100 = 125. We do the same with every other figure, putting it over £0.80 and multiplying by 100. This gives:

Year	Price	Index
1	0.80	100.00
2	1.00	125.00
3	1.10	137.50
4	1.19	148.75
5	1.23	153.75
6	1.28	160.00
7	1.32	165.00
8	1.37	171.25
9	1.44	180.00
10	1.50	187.50

2 Here we have a similar question, but you need to notice that the base year is Year 4, not Year 1:

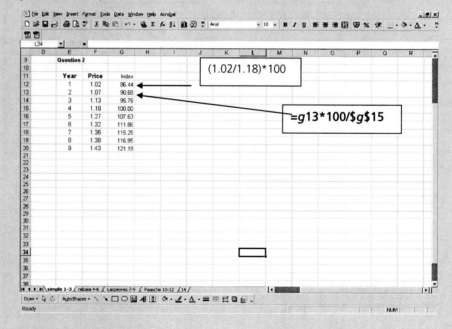

3   Although this question involves quantities, the principles are just the same – divide each number by the base year figures and multiply by 100. This gives:

Year	Quantities	Index
1	450 000	100.00
2	525,000	116.67
3	659 000	146.44
4	739 500	164.33
5	745 900	165.76
6	847 800	188.40
7	990 000	220.00
8	1 025 000	227.78

4   Whist you can do a comparison from the data as it stands, it would be much easier if the two series had the same base year. For convenience we will pick Year 1 as the common base year. This means changing the prices data. The results are shown in the screen shot below and you can see that the wages have risen by a very small amount more than the prices over the ten years. The graphs are almost identical to each other.

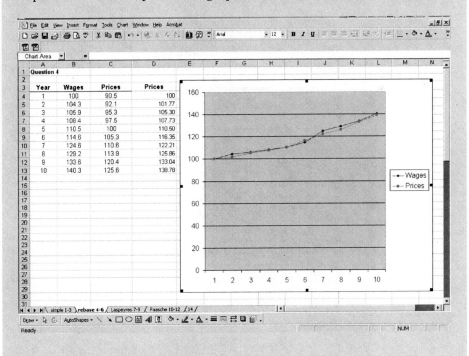

5   This question asks you to rebase the figures for productivity. Note that we use Year 2 as the base year, but after that it is simple division again.

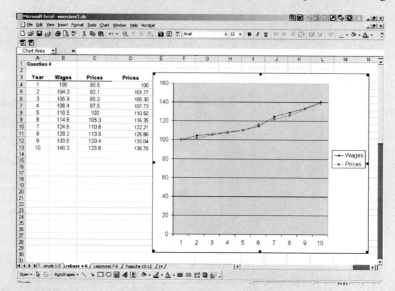

The graph allows a quick comparison of pay and productivity. You might think it better to have used Year 1 as the base year in both cases.

6   Here we need to do two sets of conversions – linking the two UK series and rebasing the US series. The screen shot below shows the results, but also includes another approach to the problem – you could work out the year-to-year changes in the data and compare these together.

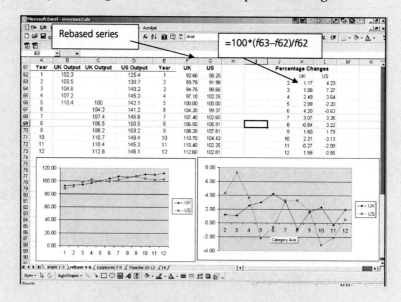

. Using percentage changes gives more manageable figures and may be easier to explain to others.

You might warn people of the possible change in definition when the UK output series was restarted in Year 5.

7 Now we reach questions which require the so called 'all items indices'. We need to combine the various prices and quantities together to get an answer. Here we are using a Laspeyres, or base-weighted index. Since there is only one set of quantity data, we can use this to weight the prices, so we will multiply each price by the appropriate weight. A spreadsheet has been used and the screen shot is shown here:

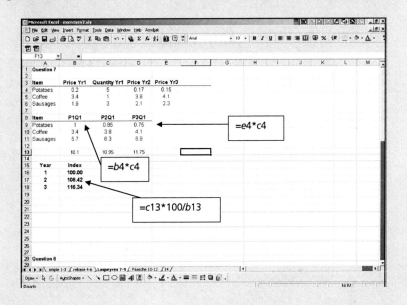

As you can see on the annotated spreadsheet, the calculations are relatively simple. The difficult part is making sure that you select the correct sets of data to multiply together.

8 We have 4 years' data but only one set of quantities, so we can only make a base-weighted index. The workings should look something like this:

Costs item	Price Yr1	Quantity Yr1	Price Yr2	Price Yr3
Coal	30	120	28	26
Ore	10	150	11	11
Scrap	2	30	2.5	3.1
Limestone	1	110	1.2	1.3
Labour	20	40	22	23

Item	P1Q1	P2Q1	P3Q1	P4Q1
Coal	3600	3360	3120	2640
Ore	1500	1650	1650	1950
Scrap	60	75	93	96
Limestone	110	132	143	165
Labour	800	880	920	960
Totals	6070	6097	5926	5811

As before, we get a set of totals from which we can calculate the index series. We get:

Year	Steel costs index	Price index
1	100.00	100.00
2	103.00	100.44
3	105.00	97.63
4	103.00	95.73

It would seem that the producers do have a case since their costs have risen whilst the prices they charge have decreased.

9 This question is very similar to the previous two and gives the following results:

Year	Index
1	100.00
2	117.27
3	136.98
4	140.49

10 This question is looking at current-weighted, or Paasche indices, since we have data on prices and quantities for each year. Again we can use a spreadsheet, but need to be even more careful that we get the correct data each time. This annotated screen shot should help to explain:

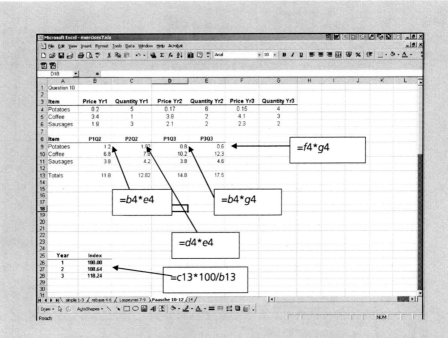

To calculate an index we need two figures, for example at C13 and B13 on the spreadsheet. They both use the Year 2 quantities as weights.

11 Here again we are looking at a company's costs and it is most appropriate to use a current-weighted index since both prices and quantities are changing over the years. Your spreadsheet may look like this:

12 A further example of current weighting giving the results:

Year	Index
1	100
2	94.4487
3	89.9757

13 Your answers will depend when you complete this exercise and which sources you consult. Some sources, such as economic trends, rebase all series to a common date, but if you go back to the individual series you will see that they start in different years.

14 Having downloaded the data, the graph is just a line graph, such as this:

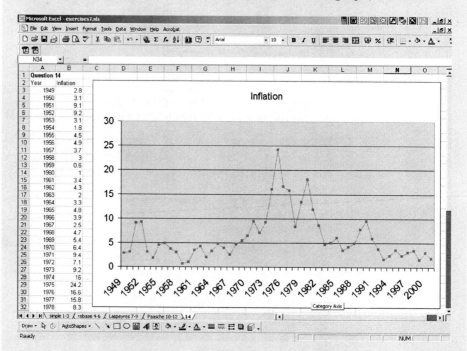

15 A quantity index uses the prices as the weights and tries to show the changes in quantities sold or purchased. We therefore need to turn the formula around to give:

$$\text{Paasche Quantity Index Year 2} = \frac{\sum (P_2 Q_2)}{\sum (P_2 Q_1)} \times 100$$

Now the labels on the prices are the same on the top and bottom, but the labels on the quantities are different. The spreadsheet will look like this:

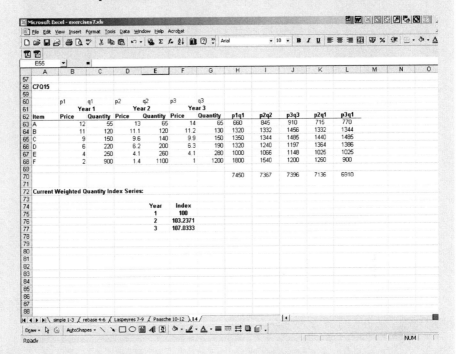

See the companion web site for further questions and annotated answers. There you will also find a PowerPoint presentation which you can use to help understand this area, or later, for revision. The site also contains links to other sites of interest.

Chapter 8

WORKING WITH POSSIBLE RELATIONSHIPS

There are a number of statistics that can be used to describe a single variable. The average number of hours worked each week by a particular group of workers (the mean) or the most frequently requested shoe size for a particular range of sports shoe (the mode) are both examples of a variable being considered on its own. However, we are often interested in how one variable relates to another. We might be interested in how the consumption of wines and spirits varies with available disposable income or how sales change when the level of advertising is changed. We first need to decide whether the data we have suggest a relationship we can work with. (It can be the case that some form of relationship does exist but that we cannot easily identify it and work with it.) We then attempt to measure the strength of the relationship, if the data allows, using the **correlation** coefficient. If the correlation coefficient supports the view that there is a relationship we can work with, we can then find the equation and use this equation to make forecasts. This chapter is only concerned with the way straight line (linear) relationships can be used to describe data.

OBJECTIVES

After reading this chapter and working through the exercises you should be able to:

- use a diagram to decide whether a possible relationship exists between two variables
- calculate the correlation coefficient, knowing its limitations
- show how a relationship can be described by an equation
- calculate the coefficients for the equation describing a simple relationship
- make a forecast using an equation
- know how the analysis could be extended using multiple regression

8.1 | PLOTTING THE DATA

The starting point is the data we have on the two variables of interest. Suppose a company is interested in the residual value of three-year-old vehicles used for lease hire. They know from experience that model, age and mileage are all important factors. It is decided to restrict the study to one make of medium-sized saloon (popular with company representatives) so that the importance of mileage alone on the residual value could be studied. Data is available on eight vehicles and is given in table 8.1 below.

Vehicle	Mileage	Residual value (£s)
1	58 379	4800
2	74 884	2990
3	18 773	8495
4	47 226	5400
5	68 454	3400
6	24 523	7200
7	62 115	4000
8	37 600	6550

TABLE 8.1   The residual value and mileage of eight selected vehicles

In this case we would expect the residual value of a vehicle to fall if a high mileage had been recorded, and we can see evidence of this in the table. To obtain a clearer picture of this relationship and also to provide a means of effective presentation we can construct a graph called a *scatter diagram*. Like all graphs we need to decide what is going to be on the vertical (*y*) axis and what is going to be on the horizontal (x) axis. The variable we are trying to explain, residual value (£s) in this case, is always plotted on the y axis. Visually, we tend to focus on the vertical axis and this is naturally our topic of interest. It is also true that others working with the graphs we construct will expect to see the variable we are trying to explain on the y axis. This variable is referred to as the **dependent variable** – its value will depend on the corresponding x value. The variable we are using for explanation is always plotted on the x axis, in this case we are using mileage, and this is referred to as the **independent variable**.

Suppose we are asked to investigate a possible relationship between crop yield and rainfall. We would expect crop yield to depend on rainfall and would therefore plot crop yield on the y axis. Rainfall is independent (within this system) and would therefore be plotted on the x axis. In this case, there is a relatively clear cause and effect and we do talk about cause and effect diagrams – our scatter diagram. Cause and effect are not always quite so clear. We might be asked to investigate the link between sales and expenditure on research and development. Most would agree that increased expenditure on research and development, which we can control, is likely to have an impact on sales. Given this dependency, we would plot sales (on the y axis) against expenditure on research and development (on the x axis). However, we need to be very careful about asserting cause and effect:

- *Relationships can be complex.* In an economic study, we might argue that sales depend on disposable income, but we could also argue that as sales increase this will impact on incomes. In this type of analysis we need to be clear about the relationship we wish to explore (we assert the relationship) and then look to the data for supportive evidence.
- *Effects can be time lagged.* If we merely plot monthly sales against monthly levels of advertising we are ignoring the fact that the impact of expenditure on advertising might be delayed.
- *Other variables might be important.* It is likely that sales will depend on a range of factors including price, delivery and quality. If we look merely at advertising, we might overlook the impact of these other factors. It would be more difficult, for example, to predict the effect on sales if we were to increase price and advertising at the same time.

Going back to our cars example (see table 8.1), the drawn scatter diagram with residual value on the y axis (£000s) and mileage on the x axis (in thousands) clearly shows the expected relationship (see figure 8.1). We can see some scattering of data points reflecting the fact that other factors can be important, like general condition and colour. The scatter diagram allows us to make a judgement as to whether the relationship can be reasonably described by a straight line, some kind of curve or appears non-existent. In this case, a straight line would give a reasonable description within our range of measurement and we would talk about a possible linear relationship.

The order of magnitude, thousands in this example, is not a problem for the software, a spreadsheet in this case, but it can be a useful simplification to work in thousands and your course may require you to do such calculations with a calculator.

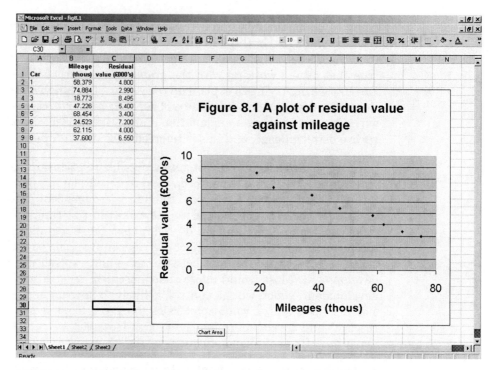

FIGURE 8.1 A scatter diagram showing residual value (£000s) plotted against mileage (in thousands)

The scatter diagram in figure 8.1 above does suggest a strong negative relationship between the two variables. It will be useful in this case to get a measure of the strength of this relationship.

8.2 | MEASURING THE STRENGTH OF A LINEAR RELATIONSHIP – THE CORRELATION COEFFICIENT

Correlation does provide a useful measure of the strength of the relationship between variables. In this case, we would like a measure of the strength of the relationship between residual value and mileage. The value we calculate is known as the *correlation coefficient* or *Pearson's correlation coefficient* and is denoted by the symbol r. This value will lie in the range $r = -1$ to $r = +1$. A value of $r = -1$ indicates a perfect negative relationship (as shown in figure 8.2) with all the data points lying on a decreasing straight line. A value of $r = +1$ indicates a perfect positive relationship (as shown in figure 8.3) with all the data points lying on an increasing straight line. It should be noted that correlation is only a measure of how well a straight line fits the data. If the data were suggestive of a curve, the low value of the correlation coefficient might be incorrectly seen as meaning no strong relationship. If we were looking at average cost, for example, the measure of correlation is likely to be low since average cost typically produces a distinctive curve, not a straight line (in these cases you would be looking to transform the data – a topic beyond this book).

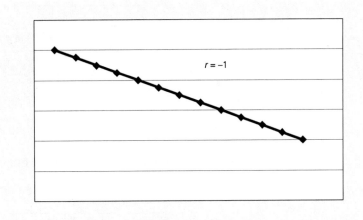

FIGURE 8.2   A perfect negative relationship

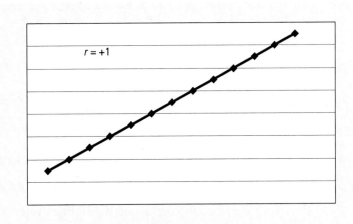

FIGURE 8.3   A perfect positive relationship

Perfect in this sense means that all the data points will lie on a straight line. Technically, we would say that the equation of the straight line would explain all the variation (differences) in the data. Given a particular value of x, we can find the value of y directly from the straight line. Such straight lines are referred to as *linear relationships*. The value of r will be closer to 0 if the scatter of points is more random, as shown in figure 8.4, or the possible relationship more non-linear, as shown in figure 8.5.

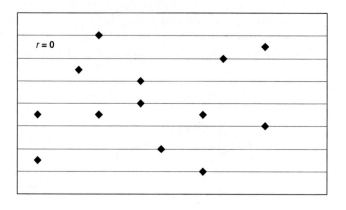

FIGURE 8.4   A random scatter of points giving a value of *r* close to 0

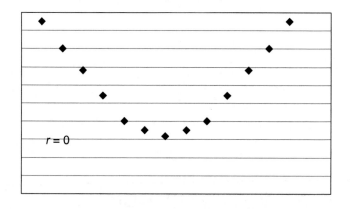

FIGURE 8.5   A non-linear relationship giving a value of *r* close to 0

Pearson's correlation coefficient, sometimes referred to as *Pearson's product moment correlation coefficient* or just the *correlation coefficient*, is calculated using the formula

$$r = \frac{n\sum xy - \sum x \sum y}{\sqrt{\left(n\sum x^2 - \left(\sum x\right)^2\right)\left(n\sum y^2 - \left(\sum y\right)^2\right)}}$$

The formula requires the following summations:

Σx	– the summation of all the x values
Σy	– the summation of all the y values
Σxy	– the multiplied values of x and y added
Σx^2	– the squared values of x added
Σy^2	– the squared values of y added
n	– the number of data points (pairs)

The spreadsheet extract, given as table 8.2, shows the various summations:

x	y	x^2	y^2	xy
58.379	4.800	3 408.108	23.040	280.219
74.884	2.990	5 607.613	8.940	223.903
18.773	8.495	352.426	72.165	159.477
47.226	5.400	2 230.295	29.160	255.020
68.454	3.400	4 685.950	11.560	232.744
24.523	7.200	601.378	51.840	176.566
62.115	4.000	3 858.273	16.000	248.460
37.600	6.550	1 413.760	42.903	246.280
391.954	42.835	22 157.803	255.608	1822.669

TABLE 8.2   The summations required for calculating the correlation and regression coefficients

The correlation coefficient can be calculated by substituting into the formula:

$$r = \frac{n\sum xy - \sum x \sum y}{\sqrt{\left(n\sum x^2 - \left(\sum x\right)^2\right)\left(n\sum y^2 - \left(\sum y\right)^2\right)}}$$

$$r = \frac{8 \times 1822.669 - 391.954 \times 42.835}{\sqrt{\left(8 \times 22\ 157.803 - (391.954)^2\right)\left(8 \times 255.608 - (42.835)^2\right)}}$$

$$r = \frac{14\ 581.352 - 16\ 789.349}{\sqrt{(177\ 262.42 - 153\ 627.93)(2044.864 - 1834.9372)}}$$

$$r = \frac{-2207.997}{\sqrt{23\ 634.49 \times 210.0268}}$$

$$r = \frac{-2207.997}{\sqrt{4\,963\,876.3}}$$

$$r = \frac{-2207.997}{2227.9758}$$

$$r = -0.9910$$

The value of $r = -0.9910$ indicates a very strong negative relationship (as expected) and suggests that further analysis is worthwhile. As a 'rule of thumb' a correlation coefficient between $r = -0.7$ and -1 would suggest a reasonably strong negative relationship and a correlation coefficient between $r = +0.7$ and $+1$ would suggest a reasonably strong positive relationship (however there is some debate on what these ranges should be and what reasonably strong actually means). The closer the value of r is to 0, the less likely any linear relationship. The squared value of r ($r^2 = 0.9821$) is called the **coefficient of determination**, and we think of this as the percentage (98 per cent in this case) of the variation explained by fitting the straight line. A more complete assessment of the correlation coefficient will require reference to a more advanced text (see Jon Curwin and Roger Slater, *Quantitative Methods for Business Decisions*, 5th ed., Thomson Learning, 2002).

Given the correlation coefficient of $r = -0.9910$, we can now consider the (straight line) equation that would describe this relationship. We show how to do this 'by hand' but you can get this directly by using the function wizard (f_x) on the Excel toolbar.

8.3 | FINDING THE EQUATION OF THE LINE – REGRESSION

The equation that describes a straight line, or linear function (see Chapter 2) takes the form

$$y = a + bx$$

where y is the dependent variable,

x is the independent variable,

a is a constant (or parameter) where the line on the graph cuts the y-axis

and

b is a constant (or parameter) giving the slope of the line

The equation could take the form

$$y = 250 + 20x$$

In this case the constant a, referred to as the **intercept**, would take the value 250, and the constant b, referred to as the **gradient**, would take the value 20. When $x = 0$, $y = 250$, which is the intercept – the point where the line on the graph cuts the y-axis. When $x = 1$, $y = 270$, showing an increase of 20 units. When $x = 2$, $y = 290$, showing a further increase of 20 units. This increase in y resulting from a unit increase in x gives the slope of the line, or gradient. A graph of this line equation is shown in figure 8.6.

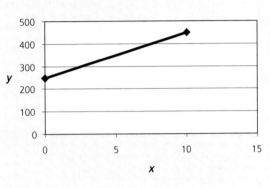

Graph showing $y = 250 + 20x$

FIGURE 8.6   A graph showing the equation $y = 250 + 20x$

Given data points on a graph, we may want to find the straight line that fits best. Even if a strong relationship exists, we would not expect all the points to lie on a straight line. We expect variation in data because of other factors or difficulties with measurement. However, if the correlation coefficient is close to –1 or +1, most points will be close to the line.

Simple linear **regression** (also referred to as line of best fit or least squares linear regression) provides a mathematical way of estimating a and b:

$$b = \frac{n\sum xy - \sum x \sum y}{n\sum x^2 - \left(\sum x\right)^2}$$

and

$$a = \frac{\sum y}{n} - b\frac{\sum x}{n}$$

Using the totals from table 8.2, we can substitute to find the values for a and b:

$$b = \frac{8 \times 1822.669 - 391.954 \times 42.835}{8 \times 22\,157.803 - \left(391.954\right)^2}$$

$$b = \frac{-2207.997}{23\,634.49}$$

$$b = -0.0934226$$

and

$$a = \frac{42.835}{8} - (-0.0934226) \times \frac{391.954}{8}$$

$$a = 5.354375 + 4.5771702$$

$$a = 9.9315452$$

The line that 'best' fits the data (making some rounding) is given by:

$$y = 9.93155 - 0.09342x$$

and is shown in figure 8.7. Here the function wizard (f_x) on the Excel toolbar is used to find intercept, slope and correlation.

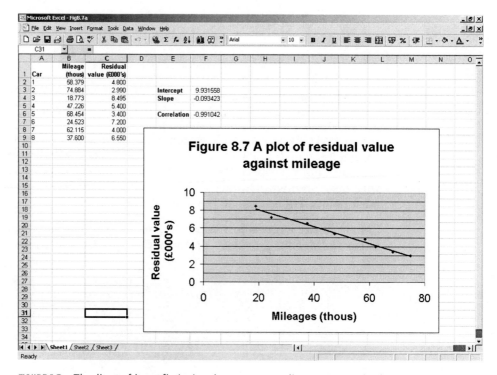

FIGURE 8.7   The line of best fit (using least squares linear regression)

8.4   MAKING A FORECAST

Now that we have established that the equation $y = 9.93155 - 0.09342x$ 'best' describes the relationship between mileage and residual value for the given data, we can use this for forecasting (or prediction) purposes.

If, for example, we were interested in the likely value of a three-year-old vehicle of this particular kind with a mileage of 45 000 miles, we could make a prediction by substitution. Let $x = 45$ (the calculations were in thousands), then:

$$y = 9.93155 - 0.09342 \times 45$$
$$= 9.93155 - 4.2039$$
$$= 5.72765$$

or £5728 (to the nearest £1).

We would feel reasonably comfortable with this forecast as it gives a residual value within the range that we have already seen. If we are making forecasts from values that lie within the range of the data, in this case x values in the range from 18 773 miles to 74 884 miles, we can be more sure about the relationship and this is referred to as **interpolation**. If however, we move outside the range of known data we cannot be so sure that the relationship still holds. Suppose we are now interested in the same sort of car with a mileage of 96 500 miles. We can again make a prediction by substitution. Let $x = 96.5$ (thousands), then:

$$y = 9.93155 - 0.09342*96.5$$
$$= 9.93155 - 9.01503$$
$$= 0.91652$$

or £917 (to the nearest £1). This value does seem rather low. The problem is that we have extended a linear relationship beyond the available data. It is often the case that over a wide range of values depreciation is more likely to follow a curve – depreciation is high to begin with and then falls at a decreasing rate. Moving outside the known range of data is referred to as **extrapolation**.

8.5   MAKING A FORECAST USING MORE THAN ONE (INDEPENDENT) VARIABLE

It is likely that the variable of interest, in this case the residual car value, is related to more than one other (independent) variable. Suppose we now have the information on 15 vehicles as shown in table 8.3.

Vehicle	Mileage	Age	Residual value (£s)
1	58 379	3	4800
2	74 884	3	2990
3	18 773	3	8495
4	47 226	3	5400
5	68 454	3	3400
6	24 523	3	7200
7	62 115	3	4000
8	37 600	3	6550
9	43 187	2	8020
10	57 445	2	8120
11	68 688	2	5100
12	28 799	4	5875
13	37 655	4	3500
14	52 038	4	3000
15	61 887	5	2000

TABLE 8.3   The residual value, age and mileage of 15 selected vehicles

We can again use an Excel spreadsheet to examine the relationship, as in figure 8.8.

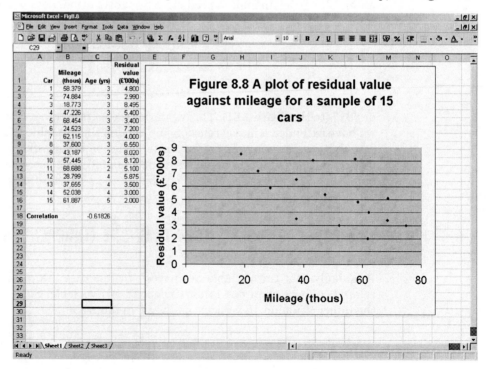

FIGURE 8.8   A plot of residual value against mileage

Now that an additional seven vehicles have been added to the data set the correlation has changed from –0.991042 to –0.61826. This still indicates a negative relationship but one that explains less of the variation in the data. The inclusion of vehicles of differing ages has added another variable that we need to take into account. It can be seen in figure 8.9 how the additional vehicles have added to the variation.

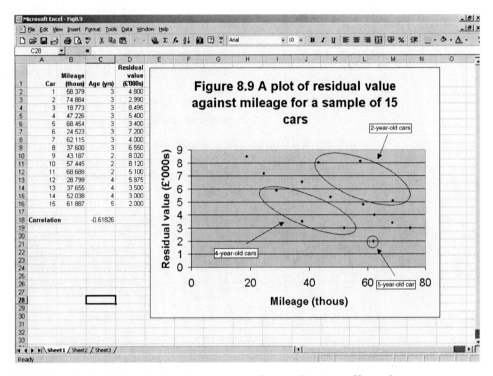

FIGURE 8.9   A plot of residual value against mileage showing effect of age

To take account of two independent (predictor) variables we can extend the linear relationship we have been working with to:

$$y = a + b_1x_1 + b_2x_2$$

where y is the dependent variable,

x_1 and x_2 are the independent variables, in this case mileage and age

a is a constant

and

b_1 and b_2 give the change in y in response to a unit change in the corresponding value of x_1 and x_2 respectively.

This form can be easily extended to three or more independent variables. The use of more than one predictor (independent) variable is referred to as **multiple regression**. A typical multiple regression printout from Excel is shown as figure 8.10 below. Such an output will give you some idea of how the subject develops but clearly we cannot discuss all the information given.

To undertake the Excel analysis you will need to load the Data Analysis add-in. The regression statistics (given as the first part of the Summary Output) give an indication of general fit (we are now beginning to get very technical):

Multiple R	(= 0.943616468) provides a measure corresponding to the correlation coefficient
R square	(= 0.890412038) the coefficient of determination
Adjusted R square	(= 0.872147378) is an unbiased estimate of the coefficient of determination (theory says R square is too large)
Standard Error	(= 0.748477578) an overall measure of variation in the regression model

To summarize this in a few words, we would say that the closeness of multiple R to 1 would suggest that this regression exercise was worthwhile and that we expect to explain 87 per cent (adjusted R^2) of the variation in the data.

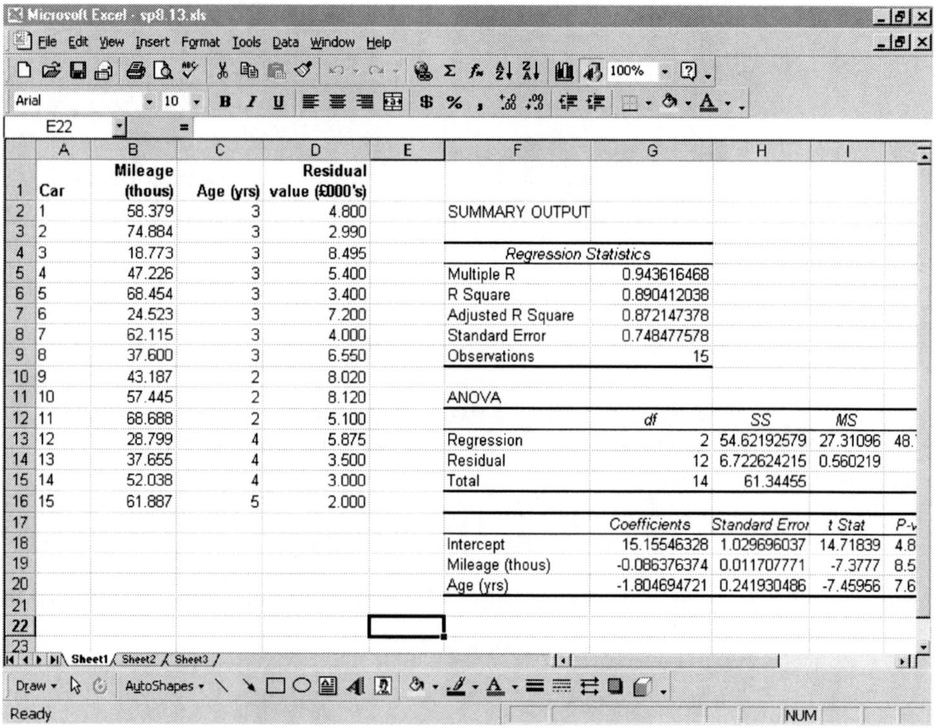

Car	Mileage (thous)	Age (yrs)	Residual value (£000's)
1	58.379	3	4.800
2	74.884	3	2.990
3	18.773	3	8.495
4	47.226	3	5.400
5	68.454	3	3.400
6	24.523	3	7.200
7	62.115	3	4.000
8	37.600	3	6.550
9	43.187	2	8.020
10	57.445	2	8.120
11	68.688	2	5.100
12	28.799	4	5.875
13	37.655	4	3.500
14	52.038	4	3.000
15	61.887	5	2.000

SUMMARY OUTPUT

Regression Statistics	
Multiple R	0.943616468
R Square	0.890412038
Adjusted R Square	0.872147378
Standard Error	0.748477578
Observations	15

ANOVA

	df	SS	MS	
Regression	2	54.62192579	27.31096	48.
Residual	12	6.722624215	0.560219	
Total	14	61.34455		

	Coefficients	Standard Error	t Stat	P-v
Intercept	15.15546328	1.029696037	14.71839	4.8
Mileage (thous)	-0.086376374	0.011707771	-7.3777	8.5
Age (yrs)	-1.804694721	0.241930486	-7.45956	7.6

FIGURE 8.10 A typical output for multiple regression

ANOVA is a more advanced topic (which we cannot consider here) and is again concerned with explaining variation.

What is of particular interest to us are the coefficients given in the last section. These allow us to form an equation (working to 3 decimal places):

$$y = 15.155 - 0.086x_1 - 1.805x_2$$

where y is the residual value (£000s)

x_1 is mileage (thousands)

and x_2 is age (years)

We can make predictions from this model (bearing in mind the same issues discussed in section 8.4). Suppose we wanted to estimate the residual value of a car that had covered 35 000 miles (35 thousand using this model) and was three years old.

By substitution:

$$y = 15.155 - 0.086 \times 35 - 1.805 \times 3 = 6.730$$

The predicted value would be 6.730 (£000s) or £6730.

It is useful to see how powerful a computer package like Excel is for statistical analysis. At one time, multiple regression would have been seen as 'out of reach' of a course like this one. The aim here is for you to know that it does exist and that using more that one predictor variable can help you deal with more complex problems.

A word of warning – including more variables can be helpful but difficulties can arise (the variables used for prediction might be highly correlated for example). If you are interested in developing your knowledge in this topic we would refer you to Jon Curwin and Roger Slater, *Quantitative Methods for Business Decisions*, 5th ed. (Thomson Learning, 2002).

8.6 │ ASSESSING THE RELATIONSHIP FOR RANKED DATA

Not all data is numeric but we may still be interested in whether an association exists between various factors. It is often possible to rank options, choices and preferences when measurement is not possible. Respondents can be asked to rank different products or services in terms of value for money, quality and other factors. To calculate Spearman's coefficient of **rank correlation** we first need to establish rank order. In these situations it is no longer possible to say how much better the first is than the second, only that the first comes before the second and so on.

Suppose we have been given the following data on the price and ranking of quality of a particular kind of power tool (table 8.4). In this case, a rank of 1 indicates highest quality.

Power tool	A	B	C	D	E	F	G	H
Price (£s)	26.99	24.99	38.50	44.99	12.99	14.99	9.99	19.99
Quality (ranking)	1	3	2	4	6	7	8	5

TABLE 8.4   The price and quality ranking of a particular power tool

The information given is a mix of numeric (price) and non-numeric (quality ranking) data. Spearman's coefficient of rank correlation can be used for mixed data sets provided we first ensure that all the data is ranked.

Spearman's coefficient of rank correlation is given by

$$r = 1 - \frac{6 \times \sum d^2}{n(n^2 - 1)}$$

Where n is the number of paired observations

and d is the difference of ranks.

The calculation of the sum of squared difference of ranks is shown in table 8.5. Note that prices have been ranked from highest to lowest.

Power tool	Price (£s)	Price rank	Quality rank	d	d²
A	26.99	3	1	2	4
B	24.99	4	3	1	1
C	38.50	2	2	0	0
D	44.99	1	4	−3	9
E	12.99	7	6	1	1
F	14.99	6	7	−1	1
G	9.99	8	8	0	0
H	19.99	5	5	0	0
					16

TABLE 8.5   Calculating of the sum of squared difference of ranks

We can now substitute in to the formula:

$$r = 1 - \frac{6 \times 16}{8(64 - 1)} = 1 - \frac{96}{504} = 1 - 0.1905 = 0.8095$$

The value of r (the rank correlation coefficient) = 0.8095 shows a high level of agreement between price and quality ranking.

8.7 | CONCLUSIONS

It is particularly useful to be able to establish a link between two or more variables. Knowing something about a possible relationship will allow us to make statements like 'sales appear to be related to the following ...' or 'the number of accidents on a particularly stretch of road seems to depend on ...' or 'the number of people taking holidays in the UK is partially explained by ...'. Correlation gives a measure of goodness of fit and regression the equation of a possible line through the data (straight lines in our examples). We have only been concerned with linear relationships but we can extend these ideas (with data transformation) to fit a variety of lines to the data. We can also extend our approach to include more than one predictor variable, using multiple regression.

The examination of the data, using graphs and measures such as correlation, is still particularly important. You must still ask the question 'does it make sense to do this'. Regression does not necessarily imply a cause and an effect. Computer software will always give you an answer. The value of the answer will depend on the quality of the data and how meaningful the analysis is that you carry out.

When you have worked through and understood the chapter you will be in a position to:

- explore relationships between variables in a variety of ways
- determine a measure of correlation and discuss the result
- apply regression analysis to a range of problems including an awareness of the use of multiple regression

8.8 EXERCISES

1 You have been given the following data on typical house price and distance from the city centre:

Distance from city centre (in kilometres)	Typical house price (£s)
5.2	89 000
5.5	87 000
6.5	98 000
7.0	112 500
8.1	120 000
9.9	145 500

a   Graph this data and comment on any possible relationship.
b   Calculate the correlation coefficient and give an interpretation to its value.
c   Determine the regression line.

d　Using the regression line found in part c, predict a typical house price for distances from the city centre of 5, 10 and 15 kilometres. Comment on your results.

2　A small business is interested in the relationship between the number of hits on its web site (measured by the number of visitors that have used the main menu) and the level of web site promotion (in £00s). Figures for the last six months are given in the table below.

Month	Web site hits	Web site promotion (£00s)
1	2500	10
2	3400	12
3	5600	16
4	5400	14
5	5500	12
6	5800	18

a　Graph the number of web site hits against web site promotion and comment on any possible relationship.

b　Calculate the correlation coefficient and give an interpretation to its value.

c　Determine the regression line.

d　Using the regression line found in part c, predict the number of web site hits if the level of monthly promotion were increased to £2000. Do you have concerns about this prediction?

e　Do you have any concerns about the simple forecasting model you have developed?

3　A personnel department now has records on ten recent employees that give an aptitude score and a job performance score, as shown below:

Employee	Aptitude score	Job performance score
A	38	26
B	59	57
C	68	34
D	40	42
E	14	23
F	33	57
G	87	83
H	71	97
I	62	72
J	81	74

a   Graph job performance score against aptitude score and comment on the relationship.

b   Determine the correlation coefficient and explain the meaning of the calculated value.

c   Determine the linear regression of job performance score against aptitude score and show this on your graph.

d   Using linear regression (your results from part c), predict the job performance score of a new employee who has just achieved a score of 50 on the aptitude test.

4   It is believed that the number of orders received by a small business are related to the number of web site enquiries and the number of telephone enquiries. The following data has been collected over the last nine weeks:

Week	Number of orders	Web site enquiries	Telephone enquiries
1	5	13	13
2	6	13	11
3	5	10	15
4	3	8	10
5	6	21	16
6	10	30	15
7	7	20	18
8	3	9	6
9	5	24	4

Present an appropriate analysis.

5   The following table gives the ranking of a group of athletes before a recent race and their positions in the race. Find an appropriate measure of correlation between the ranking and race position.

Athlete's rank	Athlete's race position
1	3
2	5
3	1
4	2
5	4
6	6
7	8
8	7

6   A personnel department now has records on ten recent employees that give an aptitude score and an interview ranking, as shown below:

Employee	Aptitude score	Interview ranking
A	38	10
B	59	5
C	68	8
D	40	7
E	14	9
F	33	1
G	87	2
H	71	3
I	62	3
J	81	6

Determine a suitable measure of correlation and comment on your result.

✓ 8.9 ANNOTATED ANSWERS

1   The screen shot below shows a graph of the relationship and the totals required for the calculations.

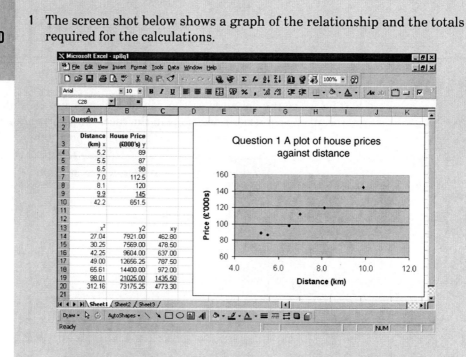

a   The graph shows a strong positive relationship and does suggest that further analysis is worthwhile.

b

$$r = \frac{6 \times 4773.30 - 42.2 \times 651.5}{\sqrt{\left(6 \times 312.6 - (42.2)^2\right)\left(6 \times 73\,175.25 - (651.5)^2\right)}}$$

$$r = \frac{1146.5}{\sqrt{92.12 \times 14\,599.25}} = 0.9886$$

The value of $r = 0.9886$ confirms a strong positive relationship.

c

$$b = \frac{6 \times 4773.30 - 42.2 \times 651.5}{6 \times 312.16 - (42.2)^2} = \frac{1146.5}{92.12} = 12.4457$$

$$a = \frac{651.5}{6} - 12.4457 \times \frac{42.2}{6} = 21.0486$$

This gives a regression equation of $y = 21.0486 + 12.4457x$. We have kept four decimal places (4dp) in this case because we have been working in thousands of £s and may want to give the answer to the nearest £1.

d

When $x = 5, y = 83.2771$ or £83 277
When $x = 10, y = 145.5056$ or £145 506
When $x = 15, y = 207.7341$ or £207 734

We feel more confident about our results if they are within the range of the data (5.2 to 9.9 kilometres), known as interpolation. In this case, none of the given x values lie within the existing range of the data but $x = 5$ and $x = 10$ would not give us any particular concerns because of their closeness. Once values are outside the known range of data, we feel less confident about using our empirical (derived from data) relationship, referred to as extrapolation. For the value $x = 15$, we would be concerned about the prediction because the radial pattern of house pricing around the city may no longer apply. At a distance of 15 kilometres you could be in a distinctly rural area or have entered another urban landscape.

2 a   The graph of web site hits against web site promotion (please note scaling of units) shows the expected positive relationship. However, there are indications that the relationship has changed with the higher level of promotion.

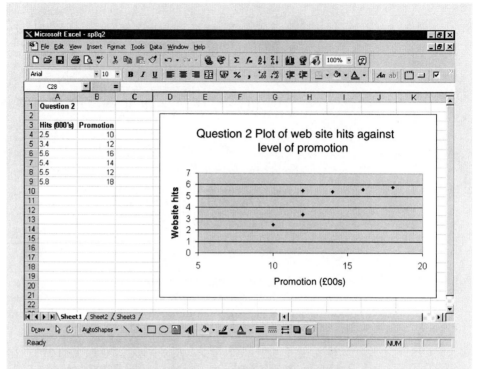

b   In this case web site hits would be the *y* variable and web site promotion the *x* variable. Given the following summary statistics (you could check these), the correlation coefficient can be found by substitution.

$$\Sigma x = 82,\ \Sigma y = 28.2,\ \Sigma y^2 = 142.22,\ \Sigma x^2 = 1164,\ \Sigma xy = 401.4$$

$$r = \frac{6 \times 401.4 - 82 \times 28.2}{\sqrt{\left(6 \times 1164 - (82)^2\right)\left(6 \times 142.22 - (28.2)^2\right)}} = \frac{96}{\sqrt{260 \times 58.08}} = 0.781216$$

c   Again by substitution, the regression coefficients can be found.

$$b = \frac{6 \times 401.4 - 82 \times 28.2}{\sqrt{\left(6 \times 1164 - (82)^2\right)}} = \frac{96}{260} = 0.369231$$

$$a = \frac{28.2}{6} - 0.369231 \times \frac{82}{6} = -0.346157$$

d   Given $y = -0.3462 + 0.3692x$ (working to 4 decimal places) by substitution (letting $x = 20$)

$$y = -0.3462 + 0.3693 \times 20 = 7.0378$$

We would expect just over 7000 (7038) hits. We are moving outside the existing range of the data (extrapolation) and would be concerned about the 'goodness of fit' of the equation.

e   this simple model (with just one predictor) only relates the number of web site hits to web site promotion. It could be the case that we would need to take account of other factors like price. The model also assumes that the effect of promotion is immediate and not time lagged in some way. We also need to be careful that the correlation is not spurious – it could be that web site interest has increased over time and the business has chosen to increase spending over time.

3   The graph, correlation and regression coefficients have been determined using a spreadsheet as shown below.

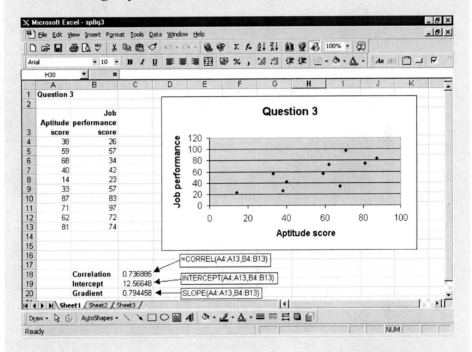

The data is sufficiently correlated to justify regression. The equation of the line is $y = 12.6 + 0.79x$. Given an aptitude score of 50 ($= x$) we would predict a job performance score of 52 (we have rounded).

4 Try to get the following output:

	A	B	C	D	E	F	G	H	I	J
1	Number of order	Website enquiries	Telephone enquiries							
2	5	13	13		SUMMARY OUTPUT					
3	6	13	11							
4	5	10	15		*Regression Statistics*					
5	3	8	10		Multiple R	0.92351393				
6	6	21	16		R Square	0.85287798				
7	10	30	15		Adjusted R Square	0.803837306				
8	7	20	18		Standard Error	0.942433598				
9	3	9	6		Observations	9				
10	5	24	4							
11					ANOVA					
12						*df*	*SS*	*MS*	*F*	*ignifica*
13					Regression	2	30.8931357	15.446568	17.39124	0.003
14					Residual	6	5.329086518	0.8881811		
15					Total	8	36.22222222			
16										
17						*Coefficients*	*Standard Error*	*t Stat*	*P-value*	*Lower 9*
18					Intercept	-0.09838898	1.060048888	-0.092816	0.929072	-2.69
19					Website enquiries	0.206069044	0.044706726	4.6093522	0.003656	0.096
20					Telephone enquiries	0.188771132	0.072763406	2.5943141	0.040974	0.010
21										
22										
23										

The results suggest that multiple regression can explain a lot of the variation in the data. Web site enquiries and telephone enquiries both positively related to the number of orders.

5

Athlete's rank	Athlete's race position	d	d²
1	3	−2	4
2	5	−3	9
3	1	2	4
4	2	2	4
5	4	1	1
6	6	0	0
7	8	−1	1
8	7	1	1
			24

$$r = 1 - \frac{6 \times 24}{8 \times 63} = 0.7143$$

6 To use Spearman's rank correlation we need to rank aptitude score first.

Aptitude score	Aptitude rank	Interview rank	d	d^2
38	8	10	−2	4
59	6	5	1	1
68	4	8	−4	16
40	7	7	0	0
14	10	9	1	1
33	9	1	8	64
87	1	2	−1	1
71	3	3	0	0
62	5	3	2	4
81	2	6	−4	16
				107

$$r = 1 - \frac{6 \times 107}{10 \times 9} = 0.3515$$

This value, closer to 0 than +1, would suggest a weak positive relationship.

WEB REFERENCE 👁

www.thomsonlearning.co.uk/businessandmanagement/curwin3

See the companion web site for further questions and annotated answers. There you will also find a PowerPoint presentation which you can use to help understand this area, or later, for revision. The site also contains links to other sites of interest.

Chapter 9

PROJECTING THE FUTURE

Much of the data we collect will be collected over time. This will give us a record of past performance and an understanding of possible trends (see Chapter 3), a **trend** being the general movement in the data over time. If we can understand past changes over time then we can consider ways of projecting these forward and using such projections as a basis for making forecasts about the future.

A lot of organizational activity is about making sense of past data. Businesses will be particularly interested in daily, weekly or monthly sales. The latest figures for price change (see section 7.7, the retail prices index) and the level of unemployment are seen as important economic indicators. Social policy will be strongly influenced by crime statistics. The demands on the health service in one winter will be seen as a guide to the demands in the next. Data collected over time is referred to as a **time series**.

In this chapter we will consider how to describe, or model, these changes over time and consider how to make reasonable projections.

OBJECTIVES

After reading this chapter and working through the exercises you should be able to:

- identify the important characteristics of data collected over time
- describe a possible trend and use linear regression, if appropriate, to model that trend
- identify a possible seasonal effect
- use a model to describe the movement in the data over time
- predict future values using a model

9.1 | USING A LINEAR (REGRESSION) MODEL

As discussed in Chapter 8, we can describe particular relationships using a straight line. Some time series data will have such a distinctive linear relationship that we can use the methods of fitting and extending a straight line.

Consider the following example. A small chain of pubs has data showing the sales of vegetarian meals over the last eight years, as given in table 9.1 below. Even if we are given the actual years (2000, 2001, 2002, 2003 and so on) it is the practice to use 1, 2, 3, 4 etc. in the calculations.

Year	1	2	3	4	5	6	7	8
Number of meals	426	399	496	588	561	588	597	648

TABLE 9.1   The number of vegetarian meals sold over the last eight years

A spreadsheet can be used to plot the number of vegetarian meals against time and to compute the straight line of 'best fit'; see figure 9.1. The values for correlation, the intercept and the gradient (slope) have been found using the function wizard (f_x) on the toolbar.

Some variation can be seen year by year but typically we are looking at increasing sales over time. The correlation coefficient of 0.9210 does suggest that fitting a straight line through the points is worthwhile. Rounding to two decimal places (2 dp) gives:

$$y = 388.25 + 33.25x$$

We could say that on average, we expect the number of vegetarian meals sold to increase by about 33 each year. If asked to forecast sales for the next three years,

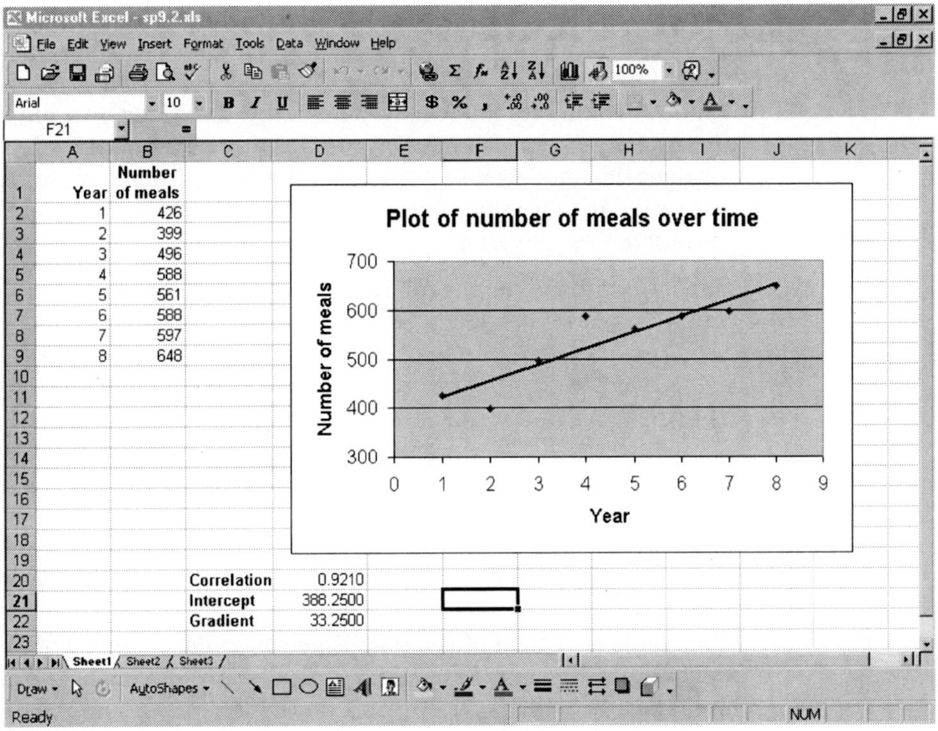

FIGURE 9.1 Sales of vegetarian meals over eight years

we would let $x = 9$, 10 and 11 (given the existing data is for x from 1 to 8 – see table 9.1). The results are given below in table 9.2.

Year	Calculation	Forecast
9	$388.25 + 33.25 \times 9$	688
10	$388.25 + 33.25 \times 10$	721
11	$388.25 + 33.25 \times 11$	754

TABLE 9.2 Forecast of sales over a three-year period

We would, of course, be very careful when working with these figures. Future sales are likely to be influenced by a range of factors including the level of economic activity and the level of competition. The figures do, however, provide a useful guide to management. If for example, this small chain of pubs actively promote vegetarian meals and future sales do indeed exceed these forecasts, then there is evidence that this promotion was worthwhile. Again, we need to be careful how we use evidence. It could be the case that sales would have increased anyway because of a more general interest in vegetarian meals. This type of projection is called *extrapolation* and was also discussed in Chapter 8.

9.2 | MODELLING A TIME SERIES

There are data sets that can be adequately described by a simple linear model. If figures increase or decrease by about the same amount each year, then fitting a straight line may well be sufficient. However, when looking at past data we are interested in more than the overall trend. We look for any regular or predictable variation.

Suppose we are given the following quarterly information on the sales of a 'learn a new language' product:

Year	Quarter 1	Quarter 2	Quarter 3	Quarter 4
1	1260	756	588	1596
2	1352	966	579	2028
3	1786	920	865	2273

TABLE 9.3   Quarterly sales of a 'learn a new language' product over the last three years

This table is typical of the way time series data is presented. The actual year may be given rather than 1, 2, 3. It is clear by looking at the columns that sales are typically higher in quarters 1 and 4 suggesting some correspondence to the academic year. Looking down the columns, sales are typically increasing (the

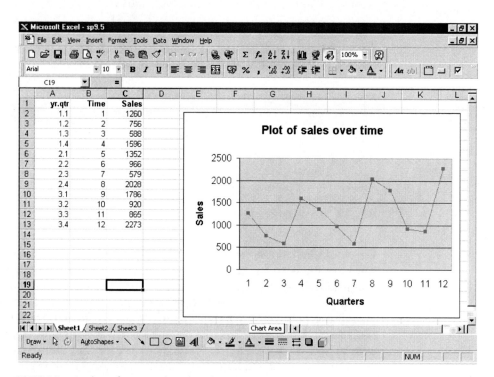

FIGURE 9.2   A plot of quarterly sales data

exceptions being quarter 3 in year 2 and quarter 2 in Year 3). A spreadsheet plot is shown as figure 9.2.

To understand the variation in the data, a model can be developed with several components. The main components are:

Trend (T) – this is the general movement in the data over time. We may expect the demand for health care to increase over time as the population ages and new medical treatments become available, for example. Over recent years we have seen increasing demand for international travel and the changes in the global travel market.

Seasonal factors (S) – these are the regular fluctuations which take place predictably in one time period, typically one year. Data is often available quarterly and we would look at the difference between spring, summer, autumn and winter. The demand for health care typically increases in the winter months and the demand for international travel typically increases in the summer months. If we are interested in retail sales, the time period could be a week, and we would look at daily variations.

Cyclical factors (C) – in a number of data sets a longer term regular variation has been observed. This is often related to economic variation. You may have come across references to the 'cycle of boom and bust'. We will not be considering this particular factor but it is useful to be aware that it does exist.

Residual or random factors (R) – these are all those factors that may make a difference at a particular point in time but we cannot include in our model. The imminence of war or unseasonally bad weather can make a difference. These random factors are usually assumed to cancel out over time but this is not always the case.

The factors can be combined in different ways to produce a model that attempts to explain the variation in the data and provide a basis for future predictions. The two models considered here are the *additive model* and the *multiplicative model*.

The additive model

In the additive model the elements of interest are added together to explain the original or actual data *(A)*:

$$A = T + S + R$$

We are not including any possible cyclical variation (C) in this model. With this model we are saying that the actual observation (A) can be explained by a general trend value (T), an allowance for any seasonal effect (S) with any difference seen as a residual or random element (R). For the model to work effectively, the residual component needs to be relatively small. To work with this model, the average value of the residual is assumed to be 0:

$$A = T + S$$

The additive model is used if the variations around the trend are of roughly the same order of magnitude. The time series given in figure 9.3 is lower in the summer quarters 2 and 3 of each year (2 and 3, 6 and 7, 10 and 11) and higher in the winter quarters by roughly the same amounts.

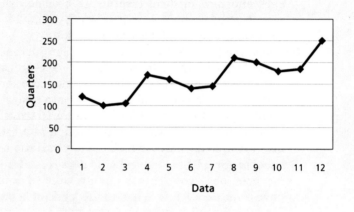

FIGURE 9.3   Time series data suggestive of the additive model

The multiplicative model

In the multiplicative model the elements of interest are multiplied together to explain the original or actual data *(A)*:

$$A = T \times S \times R$$

Again we are not including any possible cyclical variation *(C)* in this model. With this model we are saying that the actual observation *(A)* can be explained by a general trend value *(T)*, a proportionate (or percentage) increase or decrease to allow for any seasonal effect *(S)* and a proportionate (or percentage) increase or decrease to allow for any residual or random element *(R)*. In this model, we might say that sales are 10 per cent below trend or 15 per cent above trend, for example. If sales are 10 per cent below trend, then we would multiply by 0.90 and if sales are 15 per cent above trend we would multiply by 1.15 (you will see more of this later). In this model, the effects of the residual are again seen as cancelling out and the value of *R* assumed to be 1 (makes no difference when multiplied):

$$A = T \times S$$

It can be seen in figure 9.4 that the seasonal variations are increasing over time. In this example the trend is also increasing. It should be noted the seasonal effects can be increasing or decreasing over time and the trend can be increasing or decreasing over time. Such proportionate changes are common and you may well hear businesses talk about a 30 per cent increase in sales running up to the Christmas period for example.

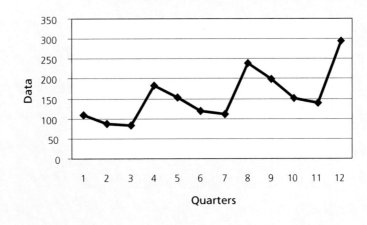

FIGURE 9.4 Time series data suggestive of the multiplicative model

We do not want to make this more complex than it needs to be. Essentially we want to work with data that has predictable variation, like seasonal differences. The two models discussed here both include the **trend** – the general movement over time. We find the trend in the same way regardless of model: decide whether the regular variation is of the roughly equal amounts (use the additive model) or is increasing or decreasing (use the multiplicative model).

9.3 | THE DETERMINATION OF TREND

Whether we use the additive model or the multiplicative model, we will still need to determine the trend. The trend is the general movement observed over time, in which we essentially smooth out the regular or predictable variations. To find the trend we use the method of **moving averages**. The values to be included in the average are taken over the natural period of variation. If we are given quarterly data (winter, spring, summer, autumn), then we would expect to include four values in the average and we would expect to see an averaging out of high and low values, as shown in table 9.4.

Yr.qtr	Time	Sales	4-point moving total	4-point moving average
1.1	1	1260		
1.2	2	756		
			4200	1050
1.3	3	588		
			4292	1073
1.4	4	1596		
			4502	1125.5
2.1	5	1352		
			4493	1123.25
2.2	6	966		
			4925	1231.25
2.3	7	579		
			5359	1339.75
2.4	8	2028		
			5313	1328.25
3.1	9	1786		
			5599	1399.75
3.2	10	920		
			5844	1461
3.3	11	865		
3.4	12	2273		

TABLE 9.4   The determination of a 4-point moving total and a 4-point moving average

It can be seen that the first four quarters are taken (winter, spring, summer, autumn) to find the first moving total of 4200. The average is found by dividing this by 4 (4200 ÷ 4). We then move on one quarter. This time we take quarters 1.2, 1.3, 1.4 and 2.1. Again all four quarters are included but in the order spring, summer, autumn and winter. The average is again found by dividing by 4.

However, there is a problem. If we are dealing with even numbers, the moving average does not align with the original values. If the natural cycle of variation is four quarters, the 4-point moving average will lie between the 2nd and 3rd quarter. Unless we align values we cannot make an easy comparison of actual values and trend values. The additional step required is to average the pairs of 4-quarter moving averages. The calculations required for a centred moving average are shown in table 9.5.

Yr.qtr	Time	Sales	4-point moving total	4-point moving average	Centred moving average
1.1	1	1260			
1.2	2	756			
			4200	1050.00	
1.3	3	588			1061.50
			4292	1073.00	
1.4	4	1596			1099.25
			4502	1125.50	
2.1	5	1352			1124.38
			4493	1123.25	
2.2	6	966			1177.25
			4925	1231.25	
2.3	7	579			1285.50
			5359	1339.75	
2.4	8	2028			1334.00
			5313	1328.25	
3.1	9	1786			1364.00
			5599	1399.75	
3.2	10	920			1430.38
			5844	1461.00	
3.3	11	865			
3.4	12	2273			

TABLE 9.5   The determination of a centred moving average

However, when presenting the spreadsheet model, we are likely to remove the empty spaces, around the results and include a plot of the trend as shown in figure 9.5. It can be seen that because of the averaging effect, the trend values start two quarters late and finish two quarters early.

To make predictions we need to extend this trend line forward. This can be done by 'eye' but given a linear type relationship we would probably use linear regression (see Chapter 8).

To determine a linear equation for the trend line we have used the values shown in table 9.6.

x	3	4	5	6	7	8	9	10
y	1062	1099	1124	1177	1286	1334	1364	1430

TABLE 9.6   Values used for trend line calculations

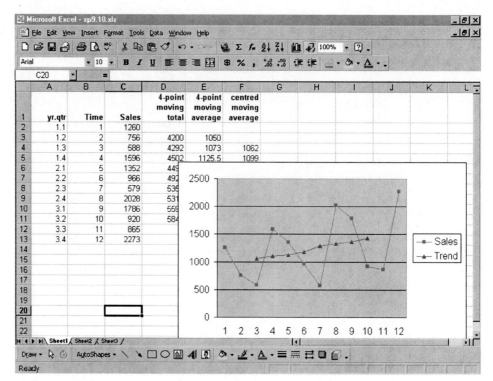

FIGURE 9.5   A typical presentation of trend calculations

The function wizard on the toolbar of Excel has been used to get a correlation coefficient of 0.9884, a slope of 55.2381 and an intercept of 875.4524. If this seems unfamiliar, refer again to Chapter 8. With a correlation coefficient so close to +1, we know that a positive straight line offers a good fit. Given the nature of the data, sales figures, we would round to whole numbers and work with the following equation:

$$y = 875 + 55x$$

To predict trend values for the coming year, quarters 1, 2, 3 and 4 of Year 4, we would use x values of 13, 14, 15 and 16. See table 9.7 below.

x	13	14	15	16
y	1590	1645	1700	1755

TABLE 9.7   The projection of trend using linear regression

To make quarterly forecasts for future years we need to do more than just project forward the trend. We also need to allow for the expected seasonal variation.

9.4 | FORECASTING ALLOWING FOR SEASONAL VARIATION

Having extended the trend forward we now need to isolate the **seasonal effects**. As discussed in section 9.2, two models are being considered – the **additive model** and the **multiplicative model**.

The additive model

This model adds the elements of interest together to explain the original or actual data (*A*):

$$A = T + S + R$$

Given that the trend values are known, the equation can be rewritten in the following form:

$$S + R = A - T$$

The residual effects are often seen as averaging out. If the residual value (*R*) can be assumed to be insignificant compared to the other effects, then

$$S = A - T$$

The seasonal effect in any one period is therefore the difference between the actual value observed and the trend value (these are the differences about the trend line that we looked at in the data plots). These differences are shown in table 9.8.

Yr.qtr	Time	Sales	4-point moving total	4-point moving average	Centred moving average	Seasonal effect (A–T)
1.1	1	1260				
1.2	2	756	4200	1050		
1.3	3	588	4292	1073	1062	–474
1.4	4	1596	4502	1125.5	1099	497
2.1	5	1352	4493	1123.25	1124	228
2.2	6	966	4925	1231.25	1177	–211
2.3	7	579	5359	1339.75	1286	–707
2.4	8	2028	5313	1328.25	1334	694
3.1	9	1786	5599	1399.75	1364	422
3.2	10	920	5844	1461.00	1430	–510
3.3	11	865				
3.4	12	2273				

TABLE 9.8   Differences between actual and trend values using the additive model

A clear pattern can be seen with the 1st and 4th quarters above trend and the 2nd and 3rd quarters below trend.

To get a best estimate of the seasonal effect, we can average these differences quarter by quarter, as shown in table 9.9.

Year	Quarter 1	Quarter 2	Quarter 3	Quarter 4
1			−474	497
2	228	−211	−707	694
3	422	−510		
Total	650	−721	−1181	1191
Average	325	−360.5	−590.5	595.5

TABLE 9.9   The averaging of seasonal differences in the additive model

This averaging of seasonal differences provides our first estimate of the seasonal effect for the additive model. However, given that any seasonal adjustment made should average out over the four quarters of the year, we would expect the sum of these averages to equal 0. In this case 325 − 360.5 − 590.5 + 595.5 = −30.5. Given that these are too low we can adjust them (make the total larger by 30.5) to make them add to 0. To do this all we need to do is divide 30.5 by 4 (30.5 ÷ 4 = 7.625) and adjust. The adjustment is shown in table 9.10 below:

Quarter	Average	Adjustment	Adjusted average	Seasonal effect (rounding)
1	325	+7.625	332.625	333
2	−360.5	+7.625	−352.875	−353
3	−590.5	+7.625	−582.875	−583
4	595.5	+7.625	603.125	603

TABLE 9.10   Adjusting the average seasonal difference

The adjusted average is the seasonal effect to be used in the additive model. After all this work we are now able to make the forecasts. The forecast is made up of the projected trend value (table 9.7) and a seasonal adjustment. See table 9.11.

Year	Quarter	Period	Trend (y)	Seasonal effect	Forecast (trend + seasonal)
4	1	13	1590	+333	1923
	2	14	1645	−353	1292
	3	15	1700	−583	1117
	4	16	1755	+603	2358

TABLE 9.11   Forecasts for Year 4 using the additive model

These figures could be added to the graph, perhaps using a different colour to show how these projections follow on from the original data.

The multiplicative model

This model multiplies the elements of interest together to explain the original or actual data *(A)*:

$$A = T \times S \times R$$

Given that the trend values are known, the equation can be rewritten in the following form:

$$S \times R = A/T$$

The residual effects are often seen as making no difference ($R = 1$ in this case) then:

$$S = A/T$$

The seasonal effect in any one period is a scaling up or down. These ratios are shown in table 9.12.

Yr.qtr	Time	Sales	4-point moving total	4-point moving average	Centred moving average	Seasonal effect (A/T)
1.1	1	1260				
1.2	2	756	4200	1050		
1.3	3	588	4292	1073	1062	0.5539
1.4	4	1596	4502	1125.5	1099	1.4519
2.1	5	1352	4493	1123.25	1124	1.2024
2.2	6	966	4925	1231.25	1177	0.8206
2.3	7	579	5359	1339.75	1286	0.4504
2.4	8	2028	5313	1328.25	1334	1.5202
3.1	9	1786	5599	1399.75	1364	1.3094
3.2	10	920	5844	1461.00	1430	0.6432
3.3	11	865				
3.4	12	2273				

TABLE 9.12 The ratio of actual and trend values using the multiplicative model

In this case, the amount that the 1st and 4th quarters are above trend and the 2nd and 3rd quarters below trend can be seen in proportionate terms e.g. the sales in the 4th quarter of Year 1 are just over 45 per cent above trend (1.4519).

To get a best estimate of the seasonal effect, we can average these ratios quarter by quarter, as shown in table 9.13.

Year	Quarter 1	Quarter 2	Quarter 3	Quarter 4
1			0.5539	1.4519
2	1.2024	0.8206	0.4504	1.5202
3	1.3094	0.6432		
Total	2.5118	1.4638	1.0043	2.9721
Average	1.2559	0.7319	0.50215	1.48605

TABLE 9.13   The averaging of proportionate differences in the multiplicative model

This averaging of proportionate differences provides our first estimate of the seasonal effect for the multiplicative model. However, given that any seasonal adjustment made should average out over the four quarters of the year, we would expect the sum of these ratios to equal 4. In this case 1.2559 + 0.7319 + 0.50215 + 1.48605 = 3.976. Given that these ratios should average out at 1 this total is too low and we need to adjust (make them add to 4). To do this all we need to do is multiply each by 4/3.976 = 1.006. The adjustment is shown in table 9.14 below:

Quarter	Average	Adjustment	Adjusted average	Seasonal effect (rounding)
1	1.2559	× 1.006	1.2634	1.26
2	0.7319	× 1.006	0.7363	0.74
3	0.50215	× 1.006	0.5052	0.51
4	1.48605	× 1.006	1.4950	1.50

TABLE 9.14   Adjusting the average proportional difference

This adjusted average is the seasonal effect to be used in the multiplicative model. We are now able to make the forecasts using the multiplicative model. The forecast is made up of the projected trend value (table 9.13) and a (multiplicative) seasonal adjustment. See table 9.15.

Year	Quarter	Period	Trend (y)	Seasonal effect	Forecast (trend × seasonal)*
4	1	13	1590	× 1.26	2003
	2	14	1645	× 0.74	1217
	3	15	1700	× 0.51	867
	4	16	1755	× 1.50	2633

*rounded to integer values

TABLE 9.15   Forecasts for Year 4 using the multiplicative model

Again forecasts could be added to the graph plot.

A comparison of models

The forecasts being made depend on the model being used. Given future uncertainties, we cannot really think about getting the right answers but we can try to identify the most appropriate model. Comparing tables 9.11 and 9.15 we can see that the multiplicative model is predicting more variability in this case (an increasing trend). This is an important point. The additive model will continue to make the same adjustment as we project forward to future years whereas the multiplicative model will make increasing or decreasing adjustments depending on whether the trend is increasing or decreasing. There are ways (statistical measures) that allow a comparison of forecasting models but these are beyond the scope of this book. What we can say in this case is that an examination of the original data (see table 9.3) would suggest variation is increasing as the trend is increasing and that we would expect the multiplicative model to perform better. Eventually, of course, we can check the accuracy of the forecast when the actual data becomes available.

9.5 | ODD NUMBER TIME PERIODS

So far we have considered a 4-point moving average and the need to 'centre' to obtain trend values. However, if the period of interest covers an odd number of recording points, like a 5-day or a 7-day working week, then there is no alignment problem and no need to 'centre'.

Suppose we are given the following data on the number of enquiries being made to a mail order business during their Monday to Friday working week. See table 9.16.

Week	Monday	Tuesday	Wednesday	Thursday	Friday
1	34	36	24	25	41
2	33	34	24	23	43
3	35	37	25	25	47

TABLE 9.16   Daily records of enquiries to a mail order business

The natural period of variation is over the five days of the working week. In this case, we would compute a 5-period moving total and a 5-period moving average as shown in table 9.17.

Week	Day	Enquiries	5-point moving total	5-point moving average
1	Monday	34		
	Tuesday	36		
	Wednesday	24	160	32.0
	Thursday	25	159	31.8
	Friday	41	157	31.4
2	Monday	33	157	31.4
	Tuesday	34	155	31.0
	Wednesday	24	157	31.4
	Thursday	23	159	31.8
	Friday	43	162	32.4
3	Monday	35	163	32.6
	Tuesday	37	165	33.0
	Wednesday	25	169	33.8
	Thursday	25		
	Friday	47		

TABLE 9.17   The determination of the 5-point moving total and 5-point moving average

As you can see, when five points are considered the moving average will align with the 3rd observation. If we were looking at a 7-day working week, then the moving average would align with the 4th observation.

The trend line can again be extended by 'eye' or using linear regression. The values used for regression are shown in table 9.18.

x	3	4	5	6	7	8	9	10	11	12	13
y	32.0	31.8	31.4	31.4	31.0	31.4	31.8	32.4	32.6	33.0	33.8

TABLE 9.18   Values used for trend line calculations

Using Excel, we get a correlation coefficient of 0.7339, a slope of 0.1836 and an intercept of 30.5855. This gives the following equation:

$$y = 30.5855 + 0.1836\,x$$

To predict trend values for the next working week (rounding to 2 decimal places), Monday to Friday in this case, we would use x values of 16, 17, 18, 19 and 20. See table 9.19 below.

x	16	17	18	19	20
y	33.52	33.71	33.89	34.07	34.26

TABLE 9.19   The projection of trend using linear regression

Having projected the trend forward we need to allow for daily variation. The effect for both the additive model $(A = T + S + R)$ and the multiplicative model $(A = T \times S \times R)$ are computed in table 9.20.

Week	Day	Enquiries	5-point moving total	5-point moving average	Seasonal effect (A–T)	Seasonal effect (A/T)
1	Monday	34				
	Tuesday	36				
	Wednesday	24	160	32.0	–8.0	0.75
	Thursday	25	159	31.8	–6.8	0.79
	Friday	41	157	31.4	9.6	1.31
2	Monday	33	157	31.4	1.6	1.05
	Tuesday	34	155	31.0	3.0	1.10
	Wednesday	24	157	31.4	–7.4	0.76
	Thursday	23	159	31.8	–8.8	0.72
	Friday	43	162	32.4	10.6	1.33
3	Monday	35	163	32.6	2.4	1.07
	Tuesday	37	165	33.0	4.0	1.12
	Wednesday	25	169	33.8	–8.8	0.74
	Thursday	25				
	Friday	47				

TABLE 9.20   A comparison of the actual and trend values using the additive and multiplicative models

To get the best estimate of the 'day of the week' effect for the additive model we average the differences, as shown in table 9.20.

Week	Monday	Tuesday	Wednesday	Thursday	Friday
1			–8.0	–6.8	9.6
2	1.6	3.0	–7.4	–8.8	10.6
3	2.4	4.0	–8.8		
Total	4.0	7.0	–24.2	–15.6	20.2
Average	2.0	3.5	–12.1*	–7.8	10.1

*this is an averaging of the three values (not two)

TABLE 9.21   The averaging of seasonal differences in the additive model

Again this averaging of seasonal differences provides our first estimate of the seasonal effect for the additive model and we would expect the sum of these averages to equal 0. In this case $2.0 + 3.5 – 12.1 – 7.8 + 10.1 = –4.3$. To adjust we need to add $4.3/5 = 0.86$ as shown in table 9.22 below:

Day	Average	Adjustment	Adjusted average
Monday	2.0	+0.86	2.86
Tuesday	3.5	+0.86	4.36
Wednesday	−12.1	+0.86	−11.24
Thursday	−7.8	+0.86	−6.94
Friday	10.1	+0.86	10.96

TABLE 9.22   Adjusting the average seasonal difference

This adjusted average is the seasonal effect to be used in the additive model. The forecast is made up of the projected trend value (table 9.19) and a seasonal adjustment. See table 9.23.

Week	Day	Period	Trend (y)	Seasonal effect	Forecast (trend + seasonal)
4	Monday	16	33.52	2.86	36.38
	Tuesday	17	33.71	4.36	38.07
	Wednesday	18	33.89	−11.24	22.65
	Thursday	19	34.07	−6.94	27.13
	Friday	20	34.26	10.96	45.22

TABLE 9.23   Forecasts for Week 4 using the additive model

To get the best estimate of the 'day of the week' effect for the multiplicative model we average the ratios as shown in table 9.24.

Week	Monday	Tuesday	Wednesday	Thursday	Friday
1			0.75	0.79	1.31
2	1.05	1.10	0.76	0.72	1.33
3	1.07	1.12	0.74		
Total	2.12	2.22	2.25	1.51	2.64
Average	1.06	1.11	0.75*	0.755	1.32

*this is an averaging of the three values (not two)

TABLE 9.24   The averaging of proportionate differences in the multiplicative model

Again this averaging of proportionate differences provides our first estimate of the seasonal effect for the multiplicative model and we would expect the sum of these ratios to equal 5. In this case $1.06 + 1.11 + 0.75 + 0.775 + 1.32 = 5.015$. To adjust we need to multiply each by $5/5.015 = 0.9970$ as shown in table 9.25:

Day	Average	Adjustment	Seasonal effect (rounded)
Monday	1.06	× 0.9970	1.06
Tuesday	1.11	× 0.9970	1.11
Wednesday	0.75	× 0.9970	0.75
Thursday	0.755	× 0.9970	0.75
Friday	1.32	× 0.9970	1.32

TABLE 9.25   Adjusting the average proportional difference

In practice this adjustment is so small as not to be significant but we have shown it for completeness.

This adjusted average is the seasonal effect to be used in the multiplicative model. The forecast is made up of the projected trend value (table 9.19) and a seasonal adjustment. See table 9.26.

Week	Day	Period	Trend (y)	Seasonal effects	Forecast (trend × seasonal)*
4	Monday	16	33.52	× 1.06	35.53
	Tuesday	17	33.71	× 1.11	37.42
	Wednesday	18	33.89	× 0.75	25.42
	Thursday	19	34.07	× 0.75	25.55
	Friday	20	34.26	× 1.32	45.22

*rounded to 2 decimal places

TABLE 9.26   Forecasts for Week 4 using the multiplicative model

Again the predictions made will depend on the model used. We would never expect predictions to turn out to be completely correct. If they were correct we would probably say this was a matter of chance. However, the use of such time series models can inform our thinking and give a guide of what to expect if other factors (e.g. the weather, the international environment) stay the same.

9.6   CONCLUSIONS

Much of the data we work with will have been collected over time. A plot against time will indicate general movement. In some cases we may be looking at a general increase or decrease that can be described by fitting a straight line, whilst in other cases we may need to consider the general trend with some predictable variation. Mathematically we attempt to explain the data using models like linear regression, the additive model and multiplicative model. These models allow us to describe what we see and make predictions about the future.

The models assume away the importance of the residual value (R). However any particular prediction will be subject to future events that we cannot know about. Heavy rain can turn to flooding or a period of pay restraint could be followed by a strike. To some extent, forecasting is about informed guessing. We use what knowledge we can, we try to give an interpretation but also need to accept that at times our forecasts are not particularly close.

However, forecasting will be important to organizations in many different ways. To manage future finance, companies will want a cash flow forecast. It is known that many new businesses will fail in the first 12 months, not because of the products or services they offer, but because of problems with cash flow. To prepare for future sales or plan stock levels companies will try to forecast demand. If these forecasts are too optimistic, a company could be left with unwanted stock, or if too pessimistic, sales opportunities will be lost. Nearly all aspects of business involve some aspect of planning for the future.

When you have worked through and understood this chapter you will be in a position to:

- discuss the use of time series data
- work with time series data in a variety of ways using appropriate models
- forecast using an appropriate time series model

9.7 EXERCISES

1 Give examples of the importance of forecasting for profit and 'not for profit' organizations.

2 Respond to the statement that 'forecasts are usually wrong'.

3 A retail organization has recorded the number of customer complaints over six months:

	March	April	May	June	July	August
Customer complaints	63	58	57	48	52	36

    a   Plot number of customer complaints against time and comment on any possible relationship.

    b   Determine the correlation coefficient and the regression of number of customer complaints against time.

    c   Use your regression to forecast the number of customer complaints in the following September and October.

    d   Suppose the actual number of customer complaints made in September was 58 and made in October was 56. How would you explain this difference from the forecasted values?

4 A company is reporting the sales of a specialist holiday package as follows:

Year	Quarter 1	Quarter 2	Quarter 3	Quarter 4
1		97	247	45
2	36	132	356	78
3	57	185	394	

a   Graph this data and comment on the outcome.
b   Use a suitable moving average to determine trend.
c   Estimate the seasonal effect using the additive and multiplicative models, and comment on your results.
d   Using linear regression or extending the trend line using a ruler, estimate the trend values for the 4th quarter of Year 3 and the 1st quarter of Year 4.
e   Forecast the sales of this specialist holiday package for the 4th quarter of Year 3 and the 1st quarter of Year 4 using the additive and multiplicative models. Comment on your results.

5 A garden centre business has recorded the number of customers each day over a 4-week period:

Week	Monday	Tuesday	Wednesday	Thursday	Friday	Saturday	Sunday
1	32	47	154	165	256	388	423
2	27	53	176	176	243	389	420
3	38	46	165	178	279	357	425
4	46	58	168	169	297	398	431

a   Graph this data and comment on the outcome.
b   Use a suitable moving average to determine trend.
c   Estimate the 'day of the week' effect using the additive and multiplicative models, and comment on your results.
d   Using linear regression or extending the trend line using a ruler, estimate the trend values for Week 5.
e   Forecast the number of customer for each day of Week 5 using the additive and multiplicative models. Comment on your results.

9.8 ANNOTATED ANSWERS

1   Trying to prepare for the future will be part of strategy for all kinds of organizations. Those organizations that need to make a profit will be particularly interested in the future demand for goods and services. They will want to extrapolate market trends and assess the impact on important business measurements like cash flow. Organizations that are 'not for profit' will also want to plan for the future. Hospitals and schools, for example, will want estimates of the level of the service they will need to provide, and will want to assess the impact on their likely budgets.

2   The skill of forecasting is to provide the best estimates for the future on the basis in current data and knowledge. It is unlikely that we will ever claim to know it all (that would soon be proved wrong!). What we can say is that we can apply models that make sense of the data and on the basis of these models we can make projections into the future. The models force us to acknowledge many of the assumptions that we make, like the importance of seasonal variation, and allow an informed discussion about the ways any forecast can be improved. The models deal with what is typical. If something is not typical, the model is not necessarily wrong, it is just not adequate for the purpose. If, for example, sales were particularly high following a revised advertising campaign, we could ask questions about both the model and the change of advertising. The important question then becomes 'what have we learnt from recent events?'

3   **a and b**   To use linear regression we let the sequential months March to August take the values 1 to 6:

Month	1	2	3	4	5	6
Customer complaints	63	58	57	48	52	36

The graph plotting (part a) and the determination of the correlation coefficient and regression coefficients (part b) are shown on the following spreadsheet:

The correlation coefficient of −0.9094 suggests a strong negative relationship that could be described by a straight line. The regression equation is $y = 68.53 − 4.63x$.

c   To forecast the number of customer complaints in September and October we would let $x = 7$ and 8.

Forecast for September ($x = 7$): $y = 68.53 − 4.63 \times 7 = 36.12$

Forecast for October ($x = 8$): $y = 68.53 − 4.63 \times 8 = 31.49$

In this case we would forecast 36 complaints in September and 31 complaints in October.

d   Regression has assumed that any established straight line relationship will just continue. Any business would welcome the falls in the number of complaints recorded in June and in August. However, there might be particular reasons for this. It might be the case that special efforts were made in those months to address the concerns or customers, or that the level of business is lower in the summer months and as a consequence the number of complaints is also lower. Whenever we undertake a forecasting exercise we need to be aware of the nature of the data. Clearly the actual numbers for September and October would give doubts about our forecasting method. We would want to know, for example, if there was a pattern of monthly differences (it might be that August is always particularly low). To improve our understanding of the number of complaints we would need this kind of monthly data over several years.

4 The quarterly variation can be seen clearly in the plot of sales over time.

Given the natural variation of four seasons, a centred 4-quarter moving average is used to determine trend (part b). The seasonal differences are also shown on the following spreadsheet for both the additive and multiplicative models.

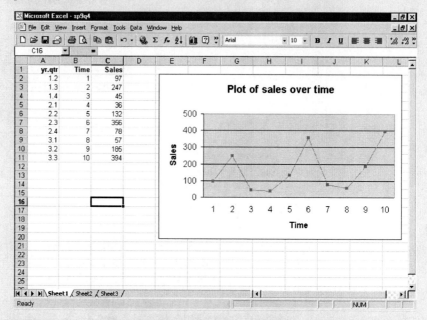

Yr.qtr	Time	Sales	4-point moving total	4-point moving average	Centred moving average	Seasonal effect (A–T)	Seasonal effect (A/T)
1.2	1	97					
1.3	2	247	425	106.25			
1.4	3	45	460	115	111	–66	0.41
2.1	4	36	569	142.25	129	–93	0.28
2.2	5	132	602	150.5	146	–14	0.90
2.3	6	356	623	155.75	153	203	2.32
2.4	7	78	676	169	162	–84	0.48
3.1	8	57	714	178.5	174	–117	0.33
3.2	9	185	636	159.00	169	16	1.10
3.3	10	394					

Having established the trend, we can identify the seasonal effects for the additive and multiplicative models (part c).

The seasonal differences for the additive model $(A - T)$:

Year	Quarter 1	Quarter 2	Quarter 3	Quarter 4
1				−66
2	−93	−14	203	−84
3	−117	16		
Total	−210	2	203	−150
Average	−105	1	*203	−75

*our single value becomes our best estimate (not ideal – we would generally want more data than this)

The sum is −105 + 1 + 203 − 75 = 24. We therefore need to adjust these averages downwards by −6 (24/4).

Quarter	Average	Adjustment	Seasonal effect
1	−105	−6	−111
2	1	−6	−5
3	203	−6	197
4	−75	−6	−81

The seasonal differences for the multiplicative model (A/T):

Year	Quarter 1	Quarter 2	Quarter 3	Quarter 4
1				0.41
2	0.28	0.90	2.32	0.48
3	0.33	1.10		
Total	0.61	2.00	2.32	0.89
Average	0.305	1.00	*2.32	0.445

* again our single value becomes our best estimate.

The sum is 0.305 + 1.00 + 2.32 + 0.445 = 4.07. We therefore need to adjust these averages downwards by 4/4.07 = 0.9828.

Quarter	Average	Adjustment	Adjusted average	Seasonal effect (rounding)
1	0.305	× 0.9828	0.2998	0.30
2	1.00	× 0.9828	0.9828	0.98
3	2.32	× 0.9828	2.2801	2.28
4	0.445	× 0.9828	0.4373	0.44

Given that the variation in the data is increasing (see graph), we would expect the multiplicative model to offer a better description.

The regression coefficients were found using:

x	3	4	5	6	7	8	9
y	111	129	146	153	162	174	169

d Using Excel, we get a correlation coefficient of 0.9561, a slope of 10.0000 and an intercept of 89.1428. This gives the following equation:

$$y = 89.1428 + 10.0000\,x$$

To estimate the trend values for the 4th quarter of Year 3 and the 1st quarter of Year 4 (rounding to 2 decimal places), we would use x values of 11 and 12.

x	11	12
y	199.14	209.14

e Forecasts using the additive and multiplicative model:

The additive model

Year	Quarter	Period	Trend (y)	Seasonal effect	Forecast (trend + seasonal)*
3	4	11	199.14	−81	118
4	1	12	209.14	−111	98

* rounded to integer values

The multiplicative model

Year	Quarter	Period	Trend (y)	Seasonal effect	Forecast (trend × seasonal)*
3	4	11	199.14	×0.44	88
4	1	12	209.14	×0.30	63

* rounded to integer values

As you can see the forecasted values from the two models are very different. In many ways only 'time will tell' which is the best. However, the historic data is suggestive of a multiplicative model. In this case there is clearly more variance being produced by the multiplicative model and we would also need to judge whether that makes business sense.

5 a The graph shows very clearly how sales increase towards the end of the week and the importance of Sunday trading for this business.

A 7-day moving average (part b) and daily differences (part c) are shown in the spreadsheet extract below:

Week	Day	Enquiries	7-day moving total	7-day moving average	Seasonal effect (A–T)	Seasonal effect (A/T)
1	Monday	32				
	Tuesday	47				
	Wednesday	154				
	Thursday	165	1465	209.3	–44.3	0.79
	Friday	256	1460	208.6	47.4	1.23
	Saturday	388	1466	209.4	178.6	1.85
	Sunday	423	1488	212.6	210.4	1.99
2	Monday	27	1499	214.1	–187.1	0.13
	Tuesday	53	1486	212.3	–159.3	0.25
	Wednesday	176	1487	212.4	–36.4	0.83
	Thursday	176	1484	212.0	–36.0	0.83
	Friday	243	1495	213.6	29.4	1.14

Continued

Week	Day	Enquiries	7-day moving total	7-day moving average	Seasonal effect (A–T)	Seasonal effect (A/T)
	Saturday	389	1488	212.6	176.4	1.83
	Sunday	420	1477	211.0	209.0	1.99
3	Monday	38	1479	211.3	–173.3	0.18
	Tuesday	46	1515	216.4	–170.4	0.21
	Wednesday	165	1483	211.9	–46.9	0.78
	Thursday	178	1488	212.6	–34.6	0.84
	Friday	279	1496	213.7	65.3	1.31
4	Saturday	357	1508	215.4	141.6	1.66
	Sunday	425	1511	215.9	209.1	1.97
	Monday	46	1502	214.6	–168.6	0.21
	Tuesday	58	1520	217.1	–159.1	0.27
	Wednesday	168	1561	223.0	–55.0	0.75
	Thursday	169	1567	223.9	–54.9	0.75
	Friday	297				
	Saturday	398				
	Sunday	431				

Having established the trend, we can identify the seasonal effects for the additive and multiplicative models (part c).

The seasonal differences for the additive model $(A - T)$:

Week	Monday	Tuesday	Wednesday	Thursday	Friday	Saturday	Sunday
1				–44.3	47.4	178.6	210.4
2	–187.1	–159.3	–36.4	–36.0	29.4	176.4	209.0
3	–173.3	–170.4	–46.9	–34.6	65.3	141.6	209.1
4	–168.6	–159.1	–55.0	–54.9			
Total	–529.0	–488.8	–138.3	–169.8	142.1	496.6	628.5
Average	–176.33	–162.93	–46.10	–42.45	47.37	165.53	209.50

The sum of these averages is –5.41. We therefore need to adjust these averages upwards by +0.77 (5.41/7).

Day	Average	Adjustment	'Day of the week' effect
Monday	−176.33	+0.77	−175.56
Tuesday	−162.93	+0.77	−162.16
Wednesday	−46.10	+0.77	−45.33
Thursday	−42.45	+0.77	−41.68
Friday	47.37	+0.77	48.14
Saturday	165.53	+0.77	166.30
Sunday	209.50	+0.77	210.27

The seasonal differences for the multiplicative model (A/T):

Week	Monday	Tuesday	Wednesday	Thursday	Friday	Saturday	Sunday
1				0.79	1.23	1.85	1.99
2	0.13	0.25	0.83	0.83	1.14	1.83	1.99
3	0.18	0.21	0.78	0.84	1.31	1.66	1.97
4	0.21	0.27	0.75	0.75			
Total	0.52	0.73	2.36	3.21	3.68	5.34	5.95
Average	0.17	0.24	0.78	0.80	1.23	1.78	1.98

The sum of these averages is 6.98. Given how close this is to the required sum of ratios (in this case) of 7 no further adjustment is necessary.

ⅾ   The regression coefficients were found using:

x	1	2	3
y	209.3	208.6	209.4

And so on, until $x = 28$.

Using Excel, we get a correlation coefficient of 0.7819, a slope of 0.4613 and an intercept of 208.4935. This gives the following equation:

$$y = 208.4935 + 0.4613x$$

To estimate the trend values for each day of Week 5 (rounding to 2 decimal places), we would use x values of 29 to 35.

Week 5	Monday	Tuesday	Wednesday	Thursday	Friday	Saturday	Sunday
x	29	30	31	32	33	34	35
y	221.87	222.33	222.79	223.26	223.72	224.18	224.64

e   Forecasts using the additive and multiplicative model:

The additive model

Forecast for Week 5

Day	Trend	'Day of the week' adjustment	Forecast*
Monday	221.87	−175.56	46
Tuesday	222.33	−162.16	60
Wednesday	222.79	−45.33	177
Thursday	223.26	−41.68	182
Friday	223.72	48.14	272
Saturday	224.18	166.30	390
Sunday	224.64	210.27	435

* rounded to integer values

The multiplicative model

Forecast for Week 5

Day	Trend	'Day of the week' adjustment	Forecast*
Monday	221.87	× 0.17	38
Tuesday	222.33	× 0.24	53
Wednesday	222.79	× 0.78	174
Thursday	223.26	× 0.80	179
Friday	223.72	× 1.23	275
Saturday	224.18	× 1.78	399
Sunday	224.64	× 1.98	445

* rounded to integer values

Again it is interesting to compare the results from the additive and multiplicative models. In this case the data is more suggestive of the additive model.

WEB REFERENCE 👁
www.thomsonlearning.co.uk/businessandmanagement/curwin3

See the companion web site for further questions and annotated answers. There you will also find a PowerPoint presentation which you can use to help understand this area, or later, for revision. The site also contains links to other sites of interest.

Chapter 10

COMPARING MONETARY RETURNS OVER TIME

When a business decides to invest for the future in some new process, marketing campaign or machinery, it wants to know if this is worthwhile. The future is uncertain, but we may be able to make some predictions about what might happen, and on the basis of these forecasts, decide whether or not to go ahead. This will not just be a case of adding up all of the benefits, since inflation will eat into the real value of future profits. We need to take into account not only what we might get, but also when we might get it.

OBJECTIVES

After reading this chapter and doing the exercises you should be able to:

- find the payback time for a project
- understand the concept of time value of money
- calculate a net present value for a project
- criticize the process
- discuss the selection of discount factors

10.1 | COSTS AND BENEFITS

The monetary aspects of a project, whether it is buying a new machine or investing in an advertising campaign, will be key to deciding whether it is worth it. However, they are not that easy to establish.

Costs are the easiest to determine, since they are usually monies that have to be paid out at the start. This means that we are fairly certain about the amounts involved. Some further costs may arise in the future and these, and their timing, will have to be estimated. Many costs can be provided by the supplier of the machine or the advertising agency which is being used.

Benefits are rather more difficult to establish. We may have projected benefits from the supplier in terms of cost savings. We may have projected sales increases from a marketing company. Each of these is uncertain, since the benefits come in the future. There may be problems of implementation, for example, some client contact software was reported recently as failing to deliver any benefits up to two years after it had been deployed with companies. (The companies had been promised considerable savings in a much shorter time.) Future sales depend on a wide variety of factors and whilst a marketing campaign may be judged a success, a downturn in the economy may mean that sales actually fall.

For these reasons, it may be better to work with a range of figures representing the future, especially in terms of likely benefits. We will develop the ideas in this chapter in terms of a single set of figures to begin with, but return later to the variety of situations discussed here.

10.2 | PAYBACK

A fairly crude method of answering the question of 'is it worth it?' is called the **payback** method. Using this method, we simply work out how long it will take to recover the initial investment. This might be found to the nearest year, month or even week. For example:

Suppose we have a company which is going to spend £100 000 on a project, and this will all be spent now. Over the next few years they expect to get back the following amounts of money:

Year	Benefits
1	£25 000
2	£50 000
3	£50 000
4	£30 000

Using the payback method, we just need to find a running total of the amount of benefit, and when this passes £100 000 we have achieved payback. In this case, table 10.1 shows that we have:

Year	Benefits	Cumulative
1	£25 000	£25 000
2	£50 000	£75 000
3	£50 000	£125 000
4	£30 000	£155 000

TABLE 10.1   A cumulative total using the payback method

As you can see, we get payback in Year 3. To be more precise, if we assume that the £50 000 comes into the company evenly through the year (i.e. a linear function), then we get to £100 000 halfway through the year. We had £75 000 at the start of the year and needed £25 000 more to reach £100 000. Since we get £50 000 during the year, the calculation is:

$$£25 000/£50 000 = ½ \text{ a year}$$

A graph could also be used to find the payback period. We would graph the cumulative benefits against time, and note when £100 000 (the cost) was passed. This is illustrated in figure 10.1 below.

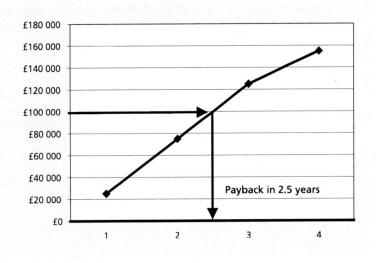

FIGURE 10.1   The payback period

This payback time might be compared to some alternative use of the funds and the one with the shortest payback period chosen as benefiting the company most. It might simply be discussed at a board meeting to ensure that payback was achieved in a 'reasonable' time. In some cases, there may be data from elsewhere in the industry which could be used as a benchmark.

Payback assumes that the future, projected benefits will be forthcoming and that any money received in the future is equivalent in value to money now. It also assumes that the money arrives in a linear fashion through the year.

10.3 | TIME-VALUE OF MONEY

This concept is fundamental to using money to assess situations or alternatives. The basic premise is that money which we receive now is worth more to us than money which we are to receive at some point in the future. This should not appear a very controversial statement for at least two reasons. Firstly, we would expect there to be some inflation in the economy, and therefore £100 now will tend to buy a larger quantity of goods than could be purchased with £100 in five or six years' time. Secondly, if we were given some money now it could be invested at some rate of interest so that in five or six years' time we would have more money.

To develop this idea further we need to look at the arithmetic associated with interest calculations. We will do this by working with an example.

Suppose that you have £100 and can invest it at 5 per cent per annum. At the end of the first year you would still have your £100 but you would also have the interest you had earned, here £5(100 × 0.05). This gives a total of £105. If this money were left invested for a second year, then you would gain further interest on the initial £100, but you would also gain interest on the £5 (here £5 × 0.05 = £0.25). At the end of the second year you would have:

$$(£100 + £5) + (£5 + £0.25) = £110.25$$

The longer you left the money invested the more interest you would gain each year, provided that the interest rate remained at 5 per cent. This is known as **compound interest**, since the interest is compounded from year to year. To develop a formula to remind us how to do these calculations we will look at the results again (table 10.2).

We use A_0 for the initial amount invested and r for the rate of interest expressed as a decimal. The amount at the end of a particular year, n, is denoted by A_n.

Year	Amount	Formula
Start, Year 0	£100	A_0
End Year 1	£100 + £5 = £105	$A_0 + A_0 \times r = A_0(1 + r)$
End Year 2	£105 + £5 + £0.25 = £110.25	$A_0(1 + r) + A_0(1 + r)r = A_0(1 + r)^2$
End Year 3	£115.7625	$A_0(1 + r)^3$
End Year n	$100(1.05)^n$	$A_0(1 + r)^n$

TABLE 10.2   Formula for compound interest calculations

There are many developments of this formula to deal with interest being paid more frequently than once per year, to allow for periodic additions to, or withdrawals from the capital sum. We will not develop these here, but if you fully understand this section, then you should have no difficulty with more complex formulae.

We can now go back to an earlier comment that money received in the future is worth less than money we have now, since today's money could be invested. By using the ideas of compound interest we can attempt to answer questions such as:

> How much is £1000 to be received in three years' time worth in today's terms if the rate of interest is 10 per cent?

It may help if we pose the question in a slightly different way:

> How much would we have to invest today at 10 per cent to get £1000 in three years' time?

Using X to represent the amount we need and putting this into our compound interest formula, we have:

$$1000 = X(1 + 0.1)^3$$

or, if we turn it around:

$$X = 1000/(1 + 0.1)^3$$
$$X = 751.32$$

This says that if we could invest £751.32 at 10 per cent for three years, we would have £1000. If this is true, then £1000 in three years' time is equivalent to £751.32 now. Given the format of our calculation, we are now in a position to work out the current value of any sum of money to be received in the future, as long as we know the rate of interest to use. The general formula would be:

$$\text{Present value} = A_n/(1 + r)^n$$

The fractional part: $\qquad\qquad 1/(1 + r)^n$

is known as the **discount factor** (or *present value factor*), and can be calculated or found from tables, e.g. Appendix 1 at the end of this book.

10.4 | NET PRESENT VALUE

This method is frequently used to assess projects and to choose between competing uses of funds. The general assumption is that the higher the **net present value (NPV)**, the better the project. In order to use NPV we need to have information about costs, benefits, timing and an interest rate to use. We will work through an example to show how to find a project NPV.

A company is trying to decide between two competing projects and has collected the following data:

Year	Project A	Project B
Cost	£120 000	£130 000
Expected contribution, Year 1	£60 000	£50 000
Expected contribution, Year 2	£65 000	£65 000
Expected contribution, Year 3	£65 000	£70 000
Expected contribution, Year 4	£60 000	£60 000
Expected contribution, Year 5	£50 000	£60 000

It uses 10 per cent as an interest or discount rate when assessing projects.

The first step in looking at this problem is to find the discount factors for each year at a rate of 10 per cent. Looking in tables or Appendix 1 (see table 10.3 here for ease of reference), these are:

Year	Discount factor
1	0.909091
2	0.826446
3	0.751315
4	0.683013
5	0.620921

TABLE 10.3   Discount factors for each year

We can now multiply each expected contribution by the appropriate discount factor (see table 10.4) to obtain the present value of that contribution.

Year	Project A	Project B		r = 10 per cent	
				PV of A	PV of B
Cost	£120 000	£130 000			
Expected contribution, Year 1	£60 000	£50 000	0.909091	54 545.45	45 454.55
Expected contribution, Year 2	£65 000	£65 000	0.826446	53 719.01	53 719.01
Expected contribution, Year 3	£65 000	£70 000	0.751315	48 835.46	52 592.04
Expected contribution, Year 4	£60 000	£60 000	0.683013	40 980.81	40 980.81
Expected contribution, Year 5	£50 000	£60 000	0.620921	31 046.07	37 255.28

TABLE 10.4   Expected contribution multiplied by the discount factor

Finally we can add up all of the present values for a project and subtract the cost. This gives the *net present value* of that project. Whilst you can do this by hand, it is much easier to use a spreadsheet. A screen shot is given as figure 10.2 below:

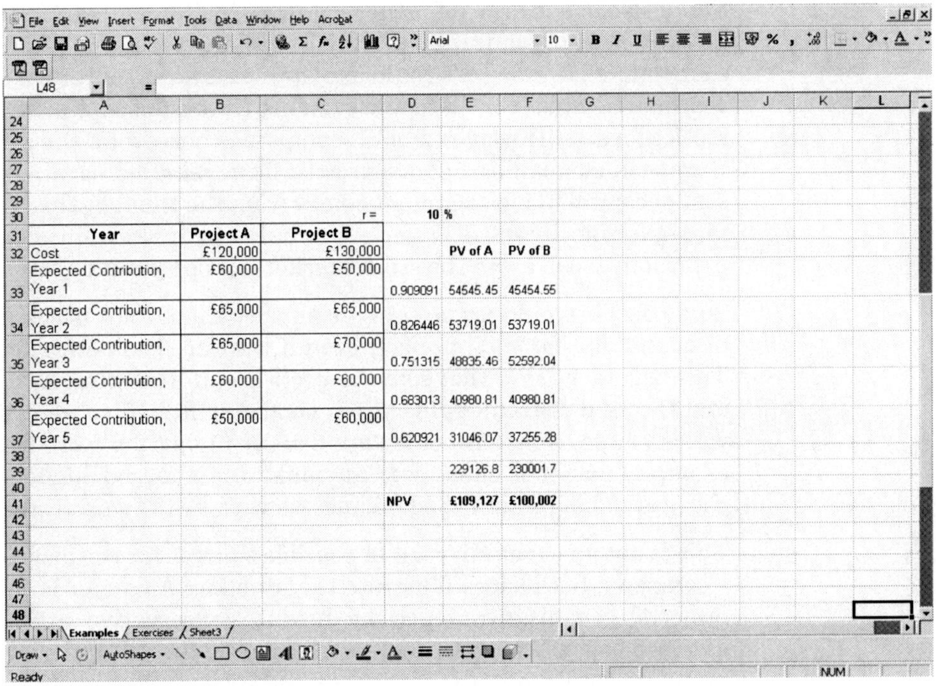

FIGURE 10.2   NPV on a spreadsheet

In this case you can see that Project A has a higher NPV than Project B. The method therefore suggests that Project A should be chosen.

10.5 | CHOOSING r

We have already discussed one of the difficulties in doing NPV calculations – that of obtaining estimates of the revenues and costs in the future. In that case, over-optimistic estimates of revenues might lead us to go ahead with a project which was not the best option for the company.

Choosing the value of r, the discount rate, could be said to give even more difficulties, since there may well be disagreement over what it is intended to measure. Perhaps it is intended to measure several things at once in some circumstances. The first thing you might take into account in choosing the discount rate is the expected level of inflation. Remember that we are trying to find the value now of money to be received in the future and this could be expressed in terms of what that money might buy. This means that we must estimate (guess) the level of inflation over the period of the project. The argument against using inflation as our value for r is that we are usually dealing with companies and the general measures of inflation (see Chapter 7) are measures of changes in retail prices.

An alternative way of choosing the discount factor would be to use a *rate of return* (a measure of business performance). Now the problem is – 'which one?'. Should we use the rate of return achieved by the company in the past? This may be the result of planning but may be simply fortuitous after particularly good economic conditions. We could use the average rate of return for the industry, if we can find what it is. We could even use the cost of capital as determined by the stock markets. However, not all companies borrow on the stock market; many European companies finance growth from retained earnings, so what is their cost of capital? This argument comes from the thought that if we didn't use the money on an internal project, we could earn a certain amount by investing externally, such as on the stock market (an opportunity cost).

Some people would argue that a company should not necessarily use the same discount rate for each project, even if they are competing for the same funds. This argument says that some projects are, inherently, more risky than others and that the discount rate chosen should reflect this risk. There may be some validity here as interest rates are usually higher for more risky borrowing, for example, personal customers at banks are charged higher rates than those charged to long established businesses with healthy profit records.

This is not an exhaustive list of possible sources for the discount rate chosen by a company, but it does illustrate the problem. A danger is that so called 'risky projects' are disadvantaged by having conservative estimates of revenues assessed by discount rates which over-estimate inflation and demand a premium return due to the risk. In such circumstances, it seems very unlikely that such a project will be chosen.

10.6 | RANGES OF BENEFITS

One way to try to overcome the problem of estimating future income streams is to make several estimates for each project. The logic here is that, whilst we have a forecast, if circumstances are particularly poor, then the estimates would be lower, and if circumstances are particularly good, then our estimates need to be higher. You might be tempted to think that this just makes the problem harder, since now we have to make several sets of estimates, and not just one. Once we decide to have more than one, then how many should we have? We often end up with three, a pessimistic one, a general one, and an optimistic one. We could just find the NPV for each set of estimates, but it would be better to find a way of combining them to get an overall estimated NPV. This can be done if we are able to assign probabilities to each set of estimates (for more on probability, see Chapter 11). The sum of the probabilities must be one.

We will illustrate this method by looking at a single project with three estimated sets of revenues.

A company is assessing a project and has three sets of projections of contribution. These are shown in the table below.

Year	Pessimistic	General	Optimistic
Cost	£100 000	£100 000	£100 000
Expected contribution, Year 1	£10 000	£12 000	£20 000
Expected contribution, Year 2	£20 000	£25 000	£40 000
Expected contribution, Year 3	£40 000	£50 000	£70 000
Expected contribution, Year 4	£25 000	£40 000	£60 000
Expected contribution, Year 5	£10 000	£20 000	£30 000

The company uses a discount rate of 12 per cent and you have determined the probabilities of the three scenarios as 0.2, 0.7 and 0.1 respectively.

The first steps are exactly as before; we find the NPV of each set of projected benefits as shown in table 10.5:

Year	Pessimistic	General	Optimistic		PV1	12 per cent PV2	PV3
Cost	£100 000	£100 000	£100 000		−£100 000	−£100 000	−£100 000
Expected contribution, Year 1	£10 000	£12 000	£20 000	0.892857	£8929	£10 714	£17 857
Expected contribution, Year 2	£20 000	£25 000	£40 000	0.797194	£15 944	£19 930	£31 888
Expected contribution, Year 3	£40 000	£50 000	£70 000	0.71178	£28 471	£35 589	£49 825
Expected contribution, Year 4	£25 000	£40 000	£60 000	0.635518	£15 888	£25 421	£38 131
Expected contribution, Year 5	£10 000	£20 000	£30 000	0.567427	£5674	£11 349	£17 023
				NPV	−£25 094	£3002	£54 723

TABLE 10.5   The NPV of each set of projected benefits

As you can see, the three scenarios give very different NPVs. In order to combine them we use the idea of **expected value**. An expected value is a number multiplied by a probability (see section 11.4 for more details). In this case we multiply each outcome by its probability and then add the three results together.

$$(-£25\ 094 \times 0.2) + (£3002 \times 0.7) + (£54\ 723 \times 0.1)$$
$$= £2555.20$$

Since this is positive, the method says that this is a worthwhile project. If we were comparing several projects, again we would be looking for the one with the highest expected NPV.

10.7 | CONCLUSIONS

Assessing future income streams will always be open to question since, by their nature, the future cash flows are uncertain. What seems obvious is that we must take into account when the money is to be received – the time value of money. Whilst the principle is obvious, doing it is more difficult conceptually since there are so many possible estimates of the future. Even if we have future cash flow projections, there still remains the problem of choosing a discount rate. Finally we may wish to also assess the probability of various outcomes. Behind all of this are other assumptions, for example, we have implicitly assumed that all benefits come at the end of the year – clearly not the case in practice.

Whilst there are limitations to this methodology it is still a very useful way of looking at projects and investments and is widely used.

When you have worked through and understood this chapter, you will be in a position to:

- explain why payback is not a very useful method of investment appraisal
- explain the time value of money
- calculate present values and net present values
- discuss the selection of discount rates
- discuss the limitations of NPV

These types of calculations are likely to be seen in any financial environment including banks and building societies. You will also see evaluations of the value of money over time in situations where a number of projects may be considered, for example in management accounting and marketing.

10.8 EXERCISES

1   A project costing £100 000 has the following expected benefits:

Year	1	2	3	4	5
Benefits	£45 000	£50 000	£50 000	£45 000	£40 000

a   If the company used the payback method, when does the project pay for itself?

b   If the company were to employ a discount rate of 10 per cent, what would be the NPV of the project?

2   A company is considering investing in a project to expand the facilities for customers. There are two different ways of doing this and they have each been costed. Projected net cash flow into the company has also been estimated.

	Cost	
Year	Project 1	Project 2
0	£120 000.00	£115 000.00
	Expected contributions	
1	£50 000.00	£40 000.00
2	£50 000.00	£45 000.00
3	£50 000.00	£50 000.00
4	£40 000.00	£50 000.00
5	£40 000.00	£45 000.00
6	£30 000.00	£30 000.00

a   If the company used the payback method, when does each project pay for itself?

b   If the company were to employ a discount rate of 12 per cent, what would be the NPV of each project?

3 Find the amount of interest earned on an investment of £2000 over three years if interest is compounded at 6 per cent.

4 You find that you are able to invest £1000. In the first year you receive 10 per cent interest. You leave the money and interest invested and receive 12 per cent in Year 2. The money and interest stays invested in Year 3, but you only get 8 per cent interest. How much money to you have at the end of Year 3?

5 A newly appointed manager decides to invest part of her salary each year to offset against possible future unemployment and, eventually, for retirement. For the first three years she is able to save £4000 per year and gets an annual interest rate (compounded) of 5 per cent. In the fourth year she is unemployed for a month, and so only invests £1000 although the interest rate is then 8 per cent. From Year 5 to Year 10 she saves £6000 per year at a compound interest rate of 7 per cent. At this point she feels confident of her employment position and decides to buy a 20-year bond with all of her savings. This pays 5 per cent per annum. How much money will she have as a lump sum if she retires when the bond matures? (Assume in each case that she invests the money on January 1st each year.)

6 A company is adding a new product range to its market offering and has made estimates of net cash flow under three assumptions: A, B and C. Use the net present value method and an interest rate of 15 per cent to determine if it is worthwhile for the company to add this product range. Note that the probability of assumptions A is assessed at 0.3, of assumptions B at 0.5, and of assumptions C at 0.2.

		Net cash flows in		
Year	Assumptions Probability	A 0.3	B 0.5	C 0.2
1		£20 000.00	£25 000.00	£35 000.00
2		£25 000.00	£35 000.00	£50 000.00
3		£30 000.00	£42 000.00	£70 000.00
4		£30 000.00	£45 000.00	£75 000.00
5		£30 000.00	£45 000.00	£78 000.00
6		£30 000.00	£45 000.00	£80 000.00

7 A new process is likely to save money for a company if it can be successfully implemented. The firm's accountants have estimated the likely savings if the implementation is fully successful, partially successful, or if it fails. The production department claim the probability of these three scenarios is 0.6, 0.3 and 0.1 respectively. The management accountants assess the probabilities as 0.3, 0.5 and 0.2 respectively.

Using the figures below together with an interest rate of 10 per cent and the NPV method, assess the expected savings over the five-year period under each set of assumptions. What would you recommend to the company?

Year		Savings	
	0.6	0.3	0.1
Probability	Fully successful	Partially successful	Fails
1	£15 000	£5000	–£10 000
2	£20 000	£8000	–£10 000
3	£25 000	£10 000	–£8000
4	£25 000	£10 000	–£8000
5	£25 000	£11 000	–£7000

N.B. Negative figures are costs

8 What factors might you take into account if you were asked to set an interest rate to be used by your company in net present value calculations?

9 A car costs £15 000 and depreciates by 20 per cent per year. What will it be worth after three years?

✔

10.9 ANNOTATED ANSWERS

1 a   If we add up the benefits we have £25 000 at the end of Year 1, £75 000 at the end of Year 2 and £125 000 at the end of Year 3. So payback is in Year 3. If we were to further assume that the money comes to the company evenly through the year, then to get to £100 000 from the end of Year 2 we need £25 000 – this is half of the benefits from Year 3, so we could say that the payback period is two and a half years.

   b   To find the NPV we can use a spreadsheet such as the one on page 230:

Spreadsheet showing:

Question 1

Year	Benefits	Cumulative		Dis Factor	PV
			r =	10 %	
1	£25,000.00	£25,000.00		0.9091	£22,727.27
2	£50,000.00	£75,000.00		0.8264	£41,322.31
3	£50,000.00	£125,000.00		0.7513	£37,565.74
4	£45,000.00	£170,000.00		0.6830	£30,735.61
5	£40,000.00	£210,000.00		0.6209	£24,836.85
			Cost		£100,000.00
			NPV		£57,187.79

=b4*e4
copy down the column

=(1/(1+(e2/100)))^a4
this can be copied down the
column

2 a   For payback we need to find the cumulative benefits and identify
when they pass the cost of the project. For both projects this will be
during Year 3:

		Cost		
Year	Project 1	Project 2	Cumulative 1	Cumulative 2
0	£120 000.00	£115 000.00		
1	£50 000.00	£40 000.00	£50 000.00	£40 000.00
2	£50 000.00	£45 000.00	£100 000.00	£85 000.00
3	£50 000.00	£50 000.00	£150 000.00	£135 000.00
4	£40 000.00	£50 000.00	£190 000.00	£185 000.00
5	£40 000.00	£45 000.00	£230 000.00	£230 000.00
6	£30 000.00	£30 000.00	£260 000.00	£260 000.00

For the first project we are £20 000 short at the beginning of the
year, and in Year 3 we get £50 000 – so we reach to total at (2/5) of
the way through the year (nearly five months). For the second
project we are £40 000 short at the beginning of the year, and earn

£50 000 in the year – so we reach the total at (4/5) of the way through the year (not quite ten months). Therefore Project 1 is better using the payback method.

b   This part is approached in much the same way as the last question, except that we apply the discount factor to each set of projected cash flows. This gives:

Discount Factor	PV1	PV2
	–£120 000.00	–£115 000.00
0.8929	£44 642.86	£35 714.29
0.7972	£44 642.86	£40 178.57
0.7118	£44 642.86	£44 642.86
0.6355	£35 714.29	£44 642.86
0.5674	£35 714.29	£40 178.57
0.5066	£26 785.71	£26 785.71
NPV =	£112 142.86	£117 142.86

Since NPV2 is larger than NPV1, Project 2 is the better project.

3   You could use a spreadsheet to work out the interest year by year:

Initial sum	£2000.00	end of Year 1	£2120.00
Interest rate	6%	end of Year 2	£2247.20
		end of Year 3	£2382.03

Each end of year gives a 6 per cent increase on the previous sum so the formula is in this format:

$$= F32 \times (1+(\$C\$32/100))$$

Or you could just apply the formula:

$$A_t = A_0 (1 + r)^t$$

Remember that 6 per cent is 0.06 as a decimal.

Giving:

$$A_3 = 2000 \times (1 + 0.06)3 = 2000 \times 1.191016 = £2382.03$$

4

Year	Initial sum	Interest rate	Interest gained	Total
1	£1000.00	10	£100.00	£1100.00
2	£1100.00	12	£132.00	£1232.00
3	£1232.00	8	£98.56	£1330.56

5 Whilst you can do this question by using the formula several times, because the sums invested and the interest rates vary over the 30-year period, it is probably easier to use a spreadsheet. This needs to calculate the position at the end of each year, up to Year 10, and then use the formula for the bond:

Year	Carried forward	Amount	Interest rate	Amount at end
1	0	£4000.00	5	£4200.00
2	4200	£4000.00	5	£8610.00
3	8610	£4000.00	5	£13 240.50
4	13 240.5	£1000.00	8	£15 37974
5	15 379.74	£6000.00	7	£22 876.32
6	22 876.322	£6000.00	7	£30 897.66
7	30 897.664	£6000.00	7	£39 480.50
8	39 480.501	£6000.00	7	£48 664.14
9	48 664.136	£6000.00	7	£58 490.63
10	58 490.625	£6,000.00	7	£69 004.97
11	69 004.969			

20-year bond investing £69 004.97 at 5 per cent p.a. compound gives £183 090.73.

There are formulae which work out amounts for compound interest with annual additions or withdrawals of capital; see Jon Curwin and Roger Slater, *Quantitative Methods for Business Decisions*, 5th ed. (Thomson Learning, 2002) for more details.

6 Here we are combining NPV calculations with probability statements. Approach such questions by doing the NPV for each set of data first, and then put in the probability factors. As before, a spreadsheet enables the calculations to be done quite quickly. Our spreadsheet is shown on the following page:

Year	Net cash flows in			r = 15%	NPVs		
	A	B	C		A	B	C
	0.3	0.5	0.2	Discount factor			
1	£20 000.00	£25 000.00	£35 000.00	0.869565217	£17 391.30	£21 739.13	£30 434.78
2	£25 000.00	£35 000.00	£50 000.00	0.756143667	£18 903.59	£26 465.03	£37 807.18
3	£30 000.00	£42 000.00	£70 000.00	0.657516232	£19 725.49	£27 615.68	£46 026.14
4	£30 000.00	£45 000.00	£75 000.00	0.571753246	£17 152.60	£25 728.90	£42 881.49
5	£30 000.00	£45 000.00	£78 000.00	0.497176735	£14 915.30	£22 372.95	£38 779.79
6	£30 000.00	£45 000.00	£80 000.00	0.432327596	£12 969.83	£19 454.74	£34 586.21
					£101 058.11	£143 376.43	£230 515.59
				Expected NPV =	£148 108.77		

The expected NPV is calculated as:

$$£101\ 058.11 \times 0.3 + £143\ 376.43 \times 0.5 + £230\ 515.59 \times 0.2 = £148\ 108.77$$

The discount factors are calculated in the usual way, but if you had to do the calculations with a calculator, then you can get these figures from sets of tables.

7 We can use the NPV calculations on savings in just the same way that we use them on net income. Our spreadsheet is shown here:

Note that we have done the sums twice, for clarity, but in practice you would just change the probability figures to move from the production department's view, to that of the management accountants.

Since both assessments lead to positive expected savings, the company would be recommended to go ahead.

8 In answering this question you can rehearse the arguments about selecting r, but you can also bring in other relevant factors.

In relation to r you might include knowing the industry the company is in, knowing the discount rate recommended by the trade body (if it exists), knowing the rate used by competitors, knowing the current Bank of England discount rate on treasury bonds, knowing the most recent economic forecasts, knowing if the project is more or less risky than other projects, knowing the discount rate used last time.

Other factors might include knowing about the predictions – are they optimistic or conservative? The assumption of success of the project. The reliability of any forecasts.

9 Depreciation is often used in accounting departments to gradually lower the value of some asset. We just take a given percentage away from the value each year. This will give us:

Year	Value	r = 20%
Start	£15 000.00	
1	£12 000.00	
2	£9 600.00	
3	£7 680.00	

WEB REFERENCE

www.thomsonlearning.co.uk/businessandmanagement/curwin3

See the companion web site for further questions and annotated answers. There you will also find a PowerPoint presentation which you can use to help understand this area, or later, for revision. The site also contains links to other sites of interest.

Chapter 11

PROBABILITY

Nothing is certain. Whatever event or outcome we might consider, there is some **chance** that it will not happen, or that a different outcome may occur. We are going to briefly look at probability as a way of trying to understand this situation. Whilst the ideas were developed from dice and card games, they are now used in insurance, forecasting, investment appraisal, portfolio management and many other areas. Your chance of surviving an operation may be assessed before you enter an operating theatre, and if it is judged too small, you may not get the operation. Since everyone differs, some subjective probability assessment must enter this judgement.

OBJECTIVES

After reading this chapter and doing the exercises you should be able to:

- define probability
- work with mutually exclusive events
- work with independent events
- work with dependent events
- calculate expected monetary values
- recognize and calculate binomial probabilities
- recognize and calculate normal distribution probabilities

235

11.1 | DEFINING PROBABILITY

This looks like a simple task at first, but a formal definition of probability is rather more difficult to find than you might think. This is a good place to start since some probability results may 'feel' wrong, or at least surprising; take for example the fact that if you have 23 people in a room, there is a chance of over 50 per cent that two of them have the same birthday and birth month.

Rather than just try to state a definition, we will try approaching the problem from several directions to understand the concept more fully. We can then develop some rules that will allow us to analyze many different probability situations. We will use dice and cards as illustrative mechanisms in some parts of this chapter.

Think about the typical example of a coin. It has two sides, heads and tails, so there are two possible outcomes to tossing the coin. If we are interested in the chance of getting a head, there is only one way of doing this, out of two possible outcomes, so we would say that the probability of getting a head is 1 divided by 2, or a half. So our definition, so far, is:

$$\text{Probability} = \frac{\text{Number of ways the event can occur}}{\text{Number of possible outcomes}}$$

Think about a die which has six sides, each with a different number 1 to 6. By this definition, the probability of a 3 is 1/6.

Most people use this definition most of the time! So why do we need to go any further? The problem is one of assumption. In order for the definition to work, we have to assume that all of the outcomes are **equally likely**. But equally likely means equally probable. Therefore we are defining probability by using probability, which doesn't make sense. (It is called a tautology.) Lots of situations allow us to get over this problem by stating that we are working with a 'fair coin' or an 'unbiased die', both of which mean that the equally likely assumption is true. The same does not necessarily apply to a real coin or die! (You must have seen or heard of a 'loaded' die.)

A second attempt might be to carry out an experiment and count how many times an event occurs – this gives us a definition:

$$\text{Probability} = \frac{\text{Number of times the event occurs}}{\text{Number of times the experiment was conducted}}$$

This is sometimes called the **frequency definition** of probability. You might guess that it will (or may) give you an answer for a particular coin or die as long as the experiment lasts long enough. It obviously won't work for a 'one-off' event.

Another approach has been to ask 'experts' what they think the probability of an event is. Then carry out some sort of averaging procedure. This **subjective probability** is very flexible, in that it will give answers for one-off or even hypothetical events, but depends on how good your 'experts' are, and how good they are at estimating probabilities.

A formal method of developing probability is through a series of axioms, but since we do not intend to develop probability theory, we shall not set off in this direction.

What does follow from these thoughts is that a probability of one means that something is certain to happen, and a probability of zero means that the event is impossible, and cannot happen.

11.2 | A FEW RULES

So far we have dealt with trying to find the probability of an event. This can be useful, but of more interest are situations where we have more than one event; each one is affected by chance. As a very simple example, think of throwing a die, and then throwing it again. The total score from the two throws is affected by chance on the first throw and chance on the second throw. The questions we need to answer are about combining probabilities of events to get an overall probability.

First we will look at events which, by their nature, cannot happen at the same time. These are usually called **mutually exclusive events**. Two such events might be getting a 3 and a 6 when a single die is thrown. The numbers appear on different sides of the die, so cannot occur at the same time. Where this is the case, we can just add the probabilities together.

> **Rule 1**: Where A and B are mutually exclusive, then:
>
> **P(A or B) = P(A) + P(B)**

The trick here is making sure that the events are really mutually exclusive.

What if the events are not mutually exclusive? The best simple example to use here is a pack of 52 playing cards. (To remind you, it has 4 suits, each with 13 cards from 2 to ace.) In this case, selecting a king does not stop you selecting, at the same time, a club, since one card, the king of clubs, does exist. Therefore the suit and the card value are not mutually exclusive. This **sample space** is illustrated in figure 11.1.

Working out this probability, we have:

$$P(King) = 4/52$$
$$P(Club) = 13/52$$

			A Clubs
K Hearts	K Diamonds	K Spades	K Clubs
			Q Clubs
			J Clubs
			10 Clubs
			9 Clubs
			8 Clubs
			7 Clubs
			6 Clubs
			5 Clubs
			4 Clubs
			3 Clubs
			2 Clubs

FIGURE 11.1   Sample space for Club or King

But each probability includes the king of clubs, so we need to take off the probability of selecting that card (1/52). So:

$$P(\text{King or Club}) = \frac{4}{52} + \frac{13}{52} - \frac{1}{52} = \frac{16}{52}$$

This leads to the second rule of probability.

Rule 2: Where A and B are not mutually exclusive, then:

P(A or B) = P(A) + P(B) – P(A and B)

This leads us on to the next type of situation: **independent events**, where the outcomes of the events are independent of each other. This means that what happens in one event has no effect whatsoever on the outcome of the other event. If you toss a coin and throw a die, getting a head on the coin does not affect what score you get on the die. Here we are interested in the chance of both events happening at the same time, and we will multiply the probabilities together. In the case given above, we have:

$$P(\text{Head}) = \frac{1}{2} \quad \text{and} \quad P(6) = 1/6$$
$$\text{So, } P(\text{Head and 6}) = (1/2) \times (1/6) = 1/12$$

This leads to another rule.

Rule 3: Where A and B are independent, then:

P(A & B) = P(A) × P(B)

Think about using two of these rules for tossing two dice and then getting the combined score. For each possible outcome, the chance is 1/36 (1/6 × 1/6 – the independence rule). This sample space is illustrated in figure 11.2. But some combinations of numbers will give the same total score, so we can add their probabilities together since they are mutually exclusive. The various scores and their probabilities are set out in figure 11.3 below where the scores are given for the first die and then the second.

Die 1

Die 2	*1*	*2*	*3*	*4*	*5*	*6*
1	2	3	4	5	6	7
2	3	4	5	6	7	8
3	4	5	6	7	8	9
4	5	6	7	8	9	10
5	6	7	8	9	10	11
6	7	8	9	10	11	12

FIGURE 11.2 Sample space for two dice

Total	Combinations	Probability
2	(1,1)	1/36
3	(1,2),(2,1)	2/36
4	(1,3),(3,1),(2,2)	3/36
5	(1,4),(4,1),(2,3),(3,2)	4/36
6	(1,5),(5,1),(2,4),(4,2),(3,3)	5/36
7	(1,6),(6,1),(2,5),(5,2),(3,4),(4,3)	6/36
8	(2,6),(6,2),(3,5),(5,3),(4,4)	5/36
9	(3,6),(6,3),(4,5),(5,4)	4/36
10	(4,6),(6,4),(5,5)	3/36
11	(5,6),(6,5)	2/36
12	(6,6)	1/36

FIGURE 11.3 Total scores on two dice

It may seem rather trivial to be working out probabilities for coins and dice, but to understand probability you need a considerable amount of practice and these things provide an easy source of such practice. Many probability situations involve the combination of these rules (and maybe dependence) with several events, and recognizing which are mutually exclusive and which are independent is key to understanding such situations.

11.3 | DEPENDENT EVENTS

Another group of events are described as **dependent**. This is the situation where the outcome of a second event is dependent upon the outcome of the first event. Again an example will help your understanding. Suppose we have a group of 11 people, three men and eight women. Now think about selecting one person, the probabilities will be:

$$P(Man) = 3/11$$
$$P(Woman) = 8/11$$

Having selected one person, we now want to select another from the remaining ten people. What are the probabilities of selecting a man or a woman? The answer is, 'it depends'. If a man was selected first, then:

$$P(Man) = 2/10$$
$$P(Woman) = 8/10$$

However, if a woman was selected first, then the probabilities are:

$$P(Man) = 3/10$$
$$P(Woman) = 7/10$$

❗ Note that if we were to select the second person from the original group (of 11 people), then the probabilities remain as they were, since the two events are then independent, and this is known as **sampling with replacement**.

In order to communicate these ideas we need some form of notation which stands for 'this is the probability of A if B happened first'. There is such a notation. We use $P(A|B)$ which can be read as 'probability of A given B'. This is sometimes referred to as conditional probability. For the example given above, we have, for the second person selected:

$$P(Man | Man) = 2/10$$
$$P(Woman | Man) = 8/10$$
$$P(Man | Woman) = 3/10$$
$$P(Woman | Woman) = 7/10$$

Dependence will always become an issue where we have selection without replacement, although, if the population is very large, say that of the UK, then the differences in the probabilities will be very, very tiny.

11.4 | EXPECTED VALUES

Whilst probability might have something to contribute to the playing of games of chance, or might put you off playing such games, it has much to contribute to decision making. If there are several possible outcomes to a decision and we can assess or assign probabilities to each, then we have a basis for deciding what to do. If we can go a step further and assign monetary values to the outcomes, then we have a decision model. (We have already used such a model in Chapter 10.)

Expected value is a useful measure of 'average' found by multiplying a set of outcomes by their probabilities.

Note: all outcomes should be included, and the probabilities should add to one.

Staying with dice, each side of the die has a probability of 1/6, so the expected score when the die is thrown will be:

$$\text{Expected Score} = \left[\frac{1}{6} \times 1 + \frac{1}{6} \times 2 + \frac{1}{6} \times 3 + \frac{1}{6} \times 4 + \frac{1}{6} \times 5 + \frac{1}{6} \times 6\right] = 3.5$$

If we have a local lottery ticket where the prize is £1000 and the chance of winning is 1/100, then the value of the ticket (i.e. the expected monetary value) is:

$$\text{EMV} = £1000 \times (1/100) + £0 \times (99/100) = £10$$

This suggests that if you were asked to pay less than £10 for the ticket, it would be worth buying on average, whilst if you were asked to pay more than £10, then it would not be worth it, on average.

11.5 | DECISION TREES

A decision tree looks at a series of choices and the likely outcomes of each. It then takes each monetary value and combines these with probabilities so that at each decision point in the process, you can express each option as an expected monetary value (EMV). The logic says that you should choose the one with the highest EMV.

Suppose you had to make a series of choices about investing a sum of money (say £100 000) that had been left to you. The first decision might be between a bank, an assurance policy and the stock market. Your return would depend upon what happened in the economy, but the effects would differ, depending on your first decision. Each of the outcomes would be in the future, so we could estimate the likely monetary value of each, and, maybe, assign probabilities to each of the outcomes. (In order to simplify this problem we will limit the likely outcomes to only a few for each case.)

Firstly we will look at the choice of a bank. If the economy is performing well, for example, then the bank may pay 5 per cent interest. If it is an average year, it pays 2 per cent, and a poor performance in the economy yields a rate of only 1 per cent. We can illustrate this by the diagram in figure 11.4.

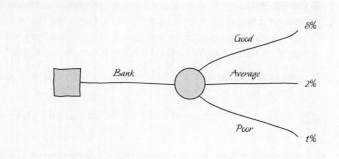

FIGURE 11.4   Outcomes if we choose a bank

Since we know how much money there is to invest, we can calculate the monetary outcome of each branch on the diagram. This would give us figure 11.5.

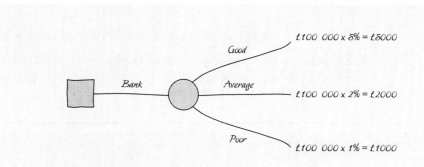

FIGURE 11.5   Monetary outcomes for the decision to use a bank

To complete this branch of the diagram, we need to know the probabilities of each outcome, here, the performance of the economy. Since we are assuming only the three outcomes, their probabilities must add to one, since something must happen. Let's say that we have an economic forecast which assesses the probabilities as follows:

$$P(\text{poor performance}) = 0.1$$

$$P(\text{average performance}) = 0.6$$

$$P(\text{good performance}) = 0.3$$

Now we can work out the EMVs for each branch by multiplying the monetary value by the probability. This is shown in figure 11.6.

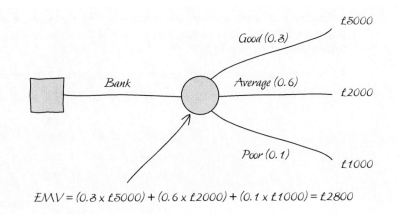

$$EMV = (0.3 \times £5000) + (0.6 \times £2000) + (0.1 \times £1000) = £2800$$

FIGURE 11.6   EMV for the bank decision

In the case of the assurance policy, you are offered a return of 4 per cent whatever happens to the economy. This branch is illustrated in figure 11.7.

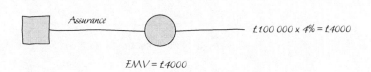

$$EMV = £4000$$

FIGURE 11.7   The assurance option

Finally we have the stock market option. Since this market is much more volatile, the returns in good times are likely to be high, but in times of poor economic performance we might expect to make losses. Let's say that in good times the return is 8 per cent, in average times the return is 3 per cent and when the

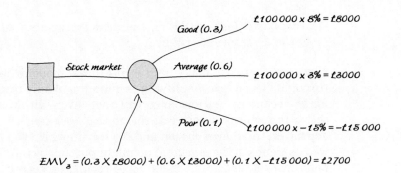

$$EMV_3 = (0.3 \times £8000) + (0.6 \times £3000) + (0.1 \times -£15\,000) = £2700$$

FIGURE 11.8   Stock market option

economy performs poorly, the return is –15 per cent. Given the probabilities suggested earlier, we can work out the EMV for the stock market option. This is illustrated in figure 11.8.

If we now combine these diagrams, as in figure 11.9, we can see that the three EMVs are:

Bank	£2800
Assurance policy	£4000
Stock market	£2700

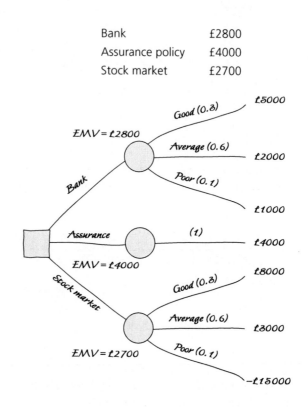

FIGURE 11.9   Decision tree

On this basis (highest EMV), the model suggests that you should invest the money in an assurance policy.

However, there are many limitations to the model which we have just worked through. We made a simplifying assumption of only three levels of performance for the economy; performance will vary, even within a year and maybe from sector to sector. We have returns on the stock market, but actual returns will depend on the shares chosen and the way in which those companies perform as well as trends in the market. You might put money into other assets, such as the bond market, or gamble it on horse racing, or keep it under the mattress! In other words, the model is useful in thinking about the problem, but has to include many simplifying assumptions. As long as we are aware of these, then we can use the model.

11.6 | BINOMIAL PROBABILITIES

Similar situations arise again and again when we are looking at probability and certain models have proved useful in understanding some situations. We will illustrate one of these probability models – the binomial distribution.

A **binomial model** will apply where:

(a) the chance of success (however defined) is constant from event to event
(b) there are two outcomes (or at least, the outcomes can be classified into just two) – success and failure

If these conditions are met, then we have a formula which will work out the probabilities. In fact, there are published tables where you can just look up the probabilities if you know the chance of success and the number of trials. Before stating the formula, we might work out the first few answers from the basic rules we learned earlier.

Let's take the situation where there are two outcomes and the chance of success remains constant from trial to trial at 0.4. Since something must happen, the chance of failure must be $(1 - 0.4) = 0.6$

If there are two trials, then there are four possible results:

1 two successes

2 success on the first trial and failure on the second

3 failure on the first trial and success on the second

4 two failures

Since the outcomes are independent, we can multiply the probabilities, so we can get the following results as shown in figure 11.10 below:

Result	Calculation	Probability
Two successes	0.4 × 0.4	0.16
Success on the first trial and failure on the second	0.4 × 0.6	0.24
Failure on the first trial and success on the second	0.6 × 0.4	0.24
Two failures	0.6 × 0.6	0.36
	Total	1.00

FIGURE 11.10   Binomial with two trials

In general, the order in which the events happen doesn't matter, so that the second and third results in figure 11.10 are seen as the same. This means that we could reduce the table to that shown in figure 11.11.

Result	Probability
2 successes	0.16
1 success	0.48
0 successes	0.36
Total	1.00

FIGURE 11.11   Reduced table

If we think of three trials with the same probabilities, then the results are:

Successes on all three trials

Success on first and second trials and failure on third

Success on first and third trials and failure on second

Success on second and third trials and failure on first

Success on first and failure on second and third trials

Success on second and failure on first and third trials

Success of third and failure on first and second trials

Failure on all three trials

Thinking about the reduced form used above, we have here:

Three successes can be achieved in only one way and this has a probability of:

$$(0.4)^3 = 0.064$$

Two successes can be achieved in three ways and has a probability each time of:

$$(0.4)^2(0.6) = 0.096$$

One success can be achieved in three ways and has a probability each time of:

$$(0.4)(0.6)^2 = 0.144$$

No success can be achieved in only one way and this has a probability of:

$$(0.6)^3 = 0.216$$

Multiplying the number of ways of achieving a result by the probability, we can get the table shown in figure 11.12.

Result	Probability
3 successes	0.064
2 successes	0.288
1 success	0.432
0 successes	0.216
Total	1.000

FIGURE 11.12   Reduced table for three trials

If we keep increasing the number of trials, the number of ways of getting a particular number of successes will need to be calculated. The formula for doing this is the **combinations** formula, which is:

$$\binom{n}{r} = \frac{n!}{r!(n-r)!}$$

This doesn't look as if it has made life any easier at first, but we still need to define what $n!$ means. It means n times one less than n, times 1 less, times 1 less, and so on until you reach 1. This is called a *factorial*. So:

$$3! = 3 \times 2 \times 1$$
$$5! = 5 \times 4 \times 3 \times 2 \times 1$$
$$10! = 10 \times 9 \times 8 \times 7 \times 6 \times 5 \times 4 \times 3 \times 2 \times 1$$
$$\text{or } 10! = 10 \times 9 \times 8 \times 7!$$

We need to define $0!$ as being equal to one.

Using the combinations formula, the number of ways to pick 2 from 3 is:

$$\binom{3}{2} = \frac{3!}{2!(3-2)!} = \frac{3 \times 2 \times 1}{2 \times 1 \times 1} = \frac{6}{2} = 3$$

This is the result we found before.

Now that we have a way of counting the number of ways of getting a particular number of successes in a certain number of trials, we can write down the binomial probability formula:

$$P(r \text{ successes in } n \text{ trials}) = \binom{n}{r} p^r (1-p)^{n-r}$$

where p is the probability of success.

As an example, think of a production line which produces thousands of screws per hour, so that the chance of finding a bent screw can be assumed to remain constant at 0.05. Find the probability of finding two or more bent screws in a sample of five taken from the production line.

Initially we need to identify the six different outcomes. These are 0, 1, 2, 3, 4 or 5 bent screws. So 'two or more' is 2, 3, 4 or 5 bent screws and we could work out each of these probabilities and then add them together (since they are mutually exclusive). Alternatively, since the probabilities of all six outcomes must add up to one, we could find the probabilities of 0 or 1 bent screws and subtract that from one. This looks easier since there are fewer calculations.

Identifying the numbers to put into the formula, we have $n = 5$, $p = 0.05$ and $r = 0$ and 1, so that:

$$P(0) = \binom{5}{0}(0.05)^0 (0.95)^5 = 0.773780938$$

$$P(1) = \binom{5}{1}(0.05)^1 (0.95)^4 = 0.203626563$$

So the probability of 0 or 1 'successes' is $0.77378 + 0.20363 = 0.97741$.

And the probability of two or more bent screws is $1 - 0.97741 = 0.02259$.

11.7 | NORMAL DISTRIBUTION

The binomial distribution is a discrete probability distribution and deals with situations where we get whole number answers. Many situations that we are likely to consider have results which can take any value, and therefore we need a continuous probability distribution. The normal distribution is the most widely used continuous probability distribution, and also forms the basis for much other work in statistics and sampling. Here we will just introduce the distribution and look at probabilities.

Normal distributions often appear where there are a large number of independent factors affecting a result or a measurement. For example, your height will depend on your age, your parents' heights, your diet, and many other factors. Heights of people in a population do, in fact, follow a normal distribution. The same is true for many 'natural' phenomena. Maybe surprisingly, it is also often true for many other phenomena such as production data and attitudes.

The first issue to consider is the shape of a normal distribution. Unlike some other distributions it always has the same basic shape, often described as a 'symmetrical bell-shaped curve'. This is shown in figure 11.13.

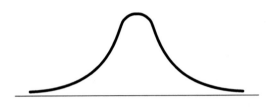

FIGURE 11.13   The normal distribution curve

Normal distributions are described by using their mean and standard deviation, (you might remember these from Chapters 5 and 6). The position of the distribution on a scale is shown by the mean, and whether it is 'tall and thin' or 'wide and flat' will be determined by the standard deviation. One thing which makes the normal distribution so useful is that every normal distribution can be related to the **standard normal distribution** by performing a simple calculation. This calculation finds the relative distance of a particular value from the mean of the distribution and uses the formula:

$$\frac{\text{Value } (X) - \text{Mean } (\mu)}{\text{Standard deviation } (\sigma)}$$

The result of this calculation is sometimes referred to as the 'z-score' for that value.

Thinking about the standard normal distribution for a moment, we can work out that the mean must be zero, since we have subtracted the original distribution mean from each value. The scale on the horizontal axis now measures the number of standard deviations from the mean, so will take values of 1, 2, 3, etc. to the right of the mean, and values of –1, –2, –3, etc. to the left of the mean. (In theory, these distributions go off to infinity in both directions, but the practical, useful part is between about plus and minus 3 standard deviations.)

The total area under the normal distribution curve is equal to one, and in probability terms, one is certainty – in other words, 'something must happen'. It follows that if we can find a particular area, this will also be a probability. This is a particularly important concept and you should just think about it for a moment. Probability is represented by the area under the curve.

Fortunately for us, the various areas under the standard normal distribution are printed in tables, as in Appendix 2, and therefore all we need to do is find the appropriate z-scores. Looking at an example will help.

> Suppose that the takings of a local shop are known to be affected by many different independent factors and follow a normal distribution with a mean of €10 000 per week and a standard deviation of €2000.
>
> (a) What is the probability of the takings in one week being over €13 000?
>
> (b) If the shopkeeper makes a profit on takings of over €9000, what is the probability of making a profit in any particular week?
>
> (c) What percentage of the time do they make a profit?

Here we have three questions, but they are all related to the normal distribution. Our first step is to identify the things which we need to know.

Here we have

the mean	= €10 000
the standard deviation	= €2 000
for part (a) the value (X)	= €13 000
for parts (b) and (c) the X value	= €9 000

A useful first step is to draw a picture of the distribution and mark the points that interest us. This is done in figure 11.14, for part (a):

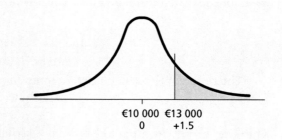

€10 000 €13 000
   0   +1.5

FIGURE 11.14 Illustrating part (a) of the shop question

The probability we are trying to find is represented by the area above the value of €13 000 and is shaded on the diagram – this shows all values above €13 000.

The next step is to convert the X value into a z-score using the formula:

$$\frac{\text{Value} - \text{Mean}}{\text{Standard deviation}} = \frac{X - \mu}{\sigma}$$

$$\frac{13\,000 - 10\,000}{2000} = 1.5$$

We can now just look up this value in the tables, as in figure 11.15.

	z	.00	.01	.02	.03	.04	.05	.06	.07
	0.0	.5000	.4960	.4920	.4880	.4840	.4801	.4761	.4721
	0.1	.4602	.4562	.4522	.4483	.4443	.4404	.4364	.4325
	0.2	.4207	.4168	.4129	.4090	.4052	.4013	.3974	.3936
	0.3	.3821	.3783	.3745	.3707	.3669	.3632	.3594	.3557
	0.4	.3446	.3409	.3372	.3336	.3300	.3264	.3228	.3192
	0.5	.3085	.3050	.3015	.2981	.2946	.2912	.2877	.2843
	0.6	.2743	.2709	.2676	.2643	.2611	.2578	.2546	.2514
	0.7	.2420	.2389	.2358	.2327	.2296	.2266	.2236	.2206
	0.8	.2119	.2090	.2061	.2033	.2005	.1977	.1949	.1922
	0.9	.1841	.1814	.1788	.1762	.1736	.1711	.1685	.1660
	1.0	.1587	.1562	.1539	.1515	.1492	.1496	.1446	.1423
	1.1	.1357	.1335	.1314	.1292	.1271	.1251	.1230	.1210
	1.2	.1151	.1132	.1112	.1093	.1075	.1056	.1038	.1020
	1.3	.0968	.0951	.0934	.0918	.0901	.0885	.0869	.0853
	1.4	.0808	.0793	.0778	.0764	.0749	.0735	.0721	.0708
z-value	1.5	.0668	.0655	.0643	.0630	.0618	.0606	.0594	.0582
	1.6	.0548	.0537	.0526	.0516	.0505	.0495	.0485	.0475
	.1.7	.0446	.0436	.0427	.0418	.0409	.0401	.0392	.0384
	.1.8	.0359	.0351	.0344	.0336	.0329	.0322	.0314	.0307
	1.9	.0287	.0381	.0274	.0268	.0267	.0256	.0250	.0244

FIGURE 11.15   Extract from normal distribution tables

As you can see, the area is 0.0668. This is the probability of takings being above €13 000 in a week.

Now turning to part (b) we can again draw a diagram to represent the situation, as in figure 11.16.

This time the value that interests us (€9000) is below the mean, and when we convert it into a z-score, we will get a negative value.

Here we have $(9000 - 10\,000)/2000 = -0.5$

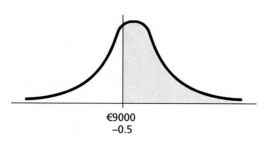

€9000
−0.5

FIGURE 11.16   Representation of part (b) of the shop question

Now the tables only give areas for positive z-scores, but we also know that the normal distribution is symmetrical, so the area to the right of a positive z-score must be identical to the area to the left of the same negative z-score. Looking at the tables (from Appendix 2) we find that the area to the right of +0.5 is 0.3085. Because of the symmetry therefore, the area to the left of −0.5 must also be 0.3085. We are interested in the area to the right of −0.5, but we know that the total area is 1, so the area to the right must be:

$$1 - 0.3085 = 0.6915$$

and this will be the probability that the shop makes a profit in a particular week.

Probabilities can be converted to percentages by multiplying by 100, so the answer to part (c) of the question is $0.6915 \times 100 = 69.15$ per cent of the time.

We should note that since we are now dealing with a continuous distribution, that we cannot find the probability of a single answer such as 'what is the probability that the shopkeeper has a turnover of exactly €10 327?' (It is possible to look at a particular value by assuming that, say 10, goes from 9.5 to 10.5 and this may help if we need to deal with discrete data, but that is beyond the scope of this chapter.) We can answer questions about the probability of values above or below a certain level, or even between two given levels. We would strongly advise you to draw quick sketches as you attempt these problems since this will help you to work out exactly which area, and hence probability, you are trying to find.

A second example. Suppose you were designing a car, but that there was very limited space for the driver. In fact you calculate that only drivers between 1.60m and 1.83m will be comfortable driving the car. Your research leads you to believe that peoples' heights are normally distributed with a mean of 1.75m and a standard deviation of 0.05m. What percentage of people will be comfortable driving your car? On this data, why might your answer be incorrect?

Figure 11.17 is a representation of this problem.

Firstly we convert the two heights into z-scores:

$$(1.60 - 1.75)/0.05 = -3$$

$$(1.83 - 1.75)/0.05 = +1.6$$

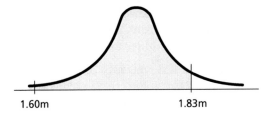

FIGURE 11.17   Area showing those between 1.60m and 1.83m

Now we can look up from the tables the area to the left of −3 = 0.00135 (remembering that it is the same as the area to the right of $z = +3$). We can also find the area to the right of $z = +1.6 = 0.0548$.

We want the area between the two heights, so we can subtract these two values from 1 (the total area under the distribution)

$$1 - 0.00135 - 0.0548 = 0.94385$$

Multiplying this answer by 100 to get a percentage means that 94.385 per cent of people would feel comfortable driving this car.

However, we have just talked about people and the average height of men and women is different, so we should really work out the percentage for each gender separately. In addition, the car might appeal to particular groups of people (a niche market) and they might not be adequately described by more general statistics. Further, if space is limited the other elements of size might be important, such as waist measurement.

If you need to take consideration to the next level, then there are some extra materials on the companion web site which should help you.

11.8 | CONCLUSIONS

If you were to develop statistics from basic principles you would need probability. This is not our aim, but probability still provides a very useful means of looking at problems. It highlights the fact that little is absolutely certain and that alternative outcomes have some chance of happening. Learning to think using probability will help you to consider areas such as contingency planning and risk analysis.

When you have worked through and understood this chapter, you will be in a position to:

- explain the concept of probability
- discuss the chance of events and combinations of events
- recognize dependence and independence

- recognize binomial situations
- calculate expected values
- use EMV in decision trees
- calculate probability for areas of a normal distribution

You will find the concepts of probability useful whenever you need to think about uncertainty, and find it typically used in an actuary's office, a management accounting department, a marketing department, a betting shop or a casino.

11.9 EXERCISES

1 What is the probability of a tail when tossing a coin?

2 What is the probability of two tails when tossing two coins?

3 What is the probability of a head followed by a tail when tossing two coins?

4 What is the probability of a head and a tail when tossing two coins?

5 When you throw a die, what is the probability of an even number?

6 If two dice are thrown, what is the probability of an even score?

7 With two dice, what is the probability of a double?

8 With a standard pack of playing cards, what is the probability of a three or a club when selecting a single card?

9 What is the probability of a red card when selecting one card from a standard deck?

10 What is the probability of an ace followed by an ace if two cards are selected without replacement?

11 What is the probability of a king or a spade on either or both of two selections from a standard pack of cards, without replacement?

12 If you have access to a spreadsheet, you can work through the 'birthdays problem'. Start off with two people in a room and calculate the probability that they have the same birthday in terms of day and month (but not year). Then calculate the answer for three people, then four, then five and so on until you get a probability of more than a half.

13 A company is about to embark on an advertising campaign costing £3m and estimates the chances of success as follows: it may be highly successful with a probability of 0.2; it may be moderately successful with a probability of 0.7; or it may have little success, which has a probability of 0.1. Successful advertising campaigns do not guarantee sales, but a consultant has calculated that a highly successful campaign will lead to very high sales with a probability of 0.6, and a moderate rise in sales with a probability of 0.4. A moderately successful campaign would lead to very high sales with a chance of 0.3, moderate increases with a

probability of 0.4, and a small increase with a probability of 0.3. Finally, a campaign which has little success may lead to moderate increases in sales with a probability of 0.3 and small increases in sales with a probability of 0.7. The accountants suggest that a very high increase in sales is worth £5m to the company; a moderate increase in sales is worth £2m; whilst small increases are only worth £1m. Draw a decision tree for this problem and calculate the expected monetary value of running the campaign. Is it worthwhile?

14 A company has just had a new, large order for its product and thinks that this may herald an expansion of the market, and therefore its sales and profits. The company could move the factory to a new and larger site which would ensure any future expansion needs were met, and this would cost £1m. The existing factory could be expanded at a cost of £0.25m, but this would limit the potential to meet particularly high increases in future market growth. Finally the new order could be met by overtime working at a cost of £0.08m. Future increases in potential sales could only be partially met by this method. Consultants suggest that there are three likely scenarios, a 40 per cent increase in sales, a 10 per cent increase in sales and a 0 per cent increase in sales, with probabilities of 0.2, 0.6 and 0.2 respectively. A new factory would easily cope with increased sales and the likely extra profit from the 40 per cent sales increase would be £6m, whereas the expanded factory would have more problems in such a large expansion and would only yield extra profits of £3.5m, whilst overtime working would be able to make only modest increases in production, thus restricting extra profits to £1.5m. If potential sales grew by 10 per cent, then a new factory would yield £2.5m extra profit, and an expanded factory £2.5m extra profit and overtime working and extra £1.5m profit. A 0 per cent increase in potential sales would lead to a zero increase in profit under any of the options. You have been asked to advise on how to react to this opportunity.

15 Evaluate: $\quad \dbinom{4}{4} ; \dbinom{4}{3} ; \dbinom{4}{2} ; \dbinom{4}{1}$ and $\dbinom{4}{0}$

16 Evaluate: $\quad \dbinom{10}{6}$ and $\dbinom{10}{1}$

17 Evaluate: $\quad \dbinom{52}{13}$

18 If 25 per cent of people have bought at least one newspaper in a particular week, for a randomly selected group of five people, find the probability that:

a   Four bought at least one paper
b   Only one bought at least one paper and

c   No-one bought a paper.

19 A multiple choice test has five answers to each question. If a candidate answers all questions randomly, what is the probability of getting four out of ten correct?

20 Many on-line visitors to a company web site just browse, but some order items. If 0.5 per cent of visitors buy from the site and there are 200 visitors per day, what is the expected number of buyers per day? What is the probability of no buyers on a day?

21 Use the tables of areas under a normal distribution (Appendix 2) to find the following areas:

a   Area to the right of $z = +1.2$
b   Area to the right of $z = +1.96$
c   Area to the right of $z = -1.55$
d   Area to the right of $z = -2.58$
e   Area to the left of $z = -2$
f   Area to the left of $z = +1.5$
g   Area between $z = +1$ and $z = +2$
h   Area between $z = -1.96$ and $z = +1.96$

22 A company sells large amounts of small value items with most invoices having about 50 different items. Errors on the invoices are known from past data to follow a normal distribution with a mean of £100 and a standard deviation of £20.

a   What is the probability an invoice is more than £125 in error?
b   What is the probability that an invoice is less than £50 in error?
c   What is the probability that an invoice is between £75 and £125 in error?

11.10 ANNOTATED ANSWERS

1 $P(T) = 0.5$, two sides of which one is a tail.

2 Only one way, so $P(2T) = 0.5 \times 0.5 = 0.25$

3 Only one way, so $P(H,T) = 0.5 \times 0.5 = 0.25$

4 Two ways – HT or TH, so, $P(H \text{ and } T) = 2 \times 0.5 \times 0.5 = 0.5$

5 Even is 2 or 4 or 6 and they are all mutually exclusive,

$$\text{so } P(\text{Even}) = (\tfrac{1}{6}) + (\tfrac{1}{6}) + (\tfrac{1}{6}) = \tfrac{1}{2}$$

6 Even is 2, 4, 6, 8, 10 or 12, but 4 and 10 can be scored in three ways each, whilst 6 and 8 can each be scored in five different ways

so, P(Even) = P(2) + P(4) + P(6) + P(8) + P(10) +P(12)
so, P(Even) = (1/36) + (3/36) + (5/36) + (5/36) + (3/36) + (1/36) = (18/36)= ½

7 Both doubles can only be scored in one way and are mutually exclusive,

so P(Double) = P(1,1) + P(2,2) + P(3,3) + P(4,4) + P(5,5) + P(6,6)
so P(Double) = (1/36) + (1/36) + (1/36) + (1/36) + (1/36) + (1/36) = (6/36) = ⅙

8 These are not mutually exclusive, so:

P(3 or Club) = P(3) + P(Club) – P(3 of Clubs)
P(3 or Club) = (4/52) + (13/52) – (1/52) = 16/52

9 Half the pack is red, so P(Red card) = ½

10 On the first card, P(Ace) = (4/52). On the second card, there are three aces left, and 51 cards, so P(Ace) = (3/51). Since we want both things to happen, we need to multiply the two probabilities together:

P(Ace,Ace) = (4/52) × (3/51) = (12/2652) = 0.0045248

11 There are several ways to get the answer to this questions; we will illustrate two of them.

Think about the first card,

P(King or Spade) = (4/52) + (13/52) – (1/52) = 16/52

This means that there are 16 cards which we could regard as successes and 36 cards which are failures. For the second card, there are only 51 left, but the number of success cards will depend on what we picked the first time.

How could we get at least one success card?

We could get success followed by failure: P(S,F) = (16/52) × (36/51)
We could get failure followed by success: (F,S) = (36/52) × (16/51)
We could get two successes: P(S,S) = (16/52) × (15/51)

All of these are mutually exclusive, so we can add the probabilities together. The answer is (116/221).

An alternative way to do the problem is to say 'How could we fail'? There is only one way, fail on first card and fail on second:

P(F,F) = (36/52) × (35/51)

We want to succeed, so we subtract this probability from 1

P(at least 1 Success) = 1 – P(F,F) = 1 – {(36/52) × (35/51)} = (116/221)

12 This is another problem where it is easier to work out the probability of something not happening, rather than the probability it does happen, since there are lots of different ways to succeed, but few to fail.

For 2 people, they have different birthdays if the second person has their birthday on one of the 364 days which is not the first person's birthday. (Note that we are simplifying the problem by assuming only 365 days per year.) So, the chance that they have different birthdays is: (364/365)

and the chance they have the same birthday must be 1 − P(different birthdays): 1 − (364/365) = 0.0027398

For 3 people, the third person must have a birthday which is different from each of the others if we are to fall, on one of the 363 other days. We need this to happen as well as the first case, so we multiply the probabilities. Again we subtract the result from 1:

P(Success) = 1 − (364/365) × (363/365) = 0.0082042

We suggest that you use a spreadsheet to calculate the remaining probabilities. You should reach a probability of over 0.5 when you have 23 people in the room.

13 The diagram will have three main branches for high, moderate and low success in the campaign and each will be followed by sales outcomes, with probabilities and then finally, the profit outcomes. This is shown below:

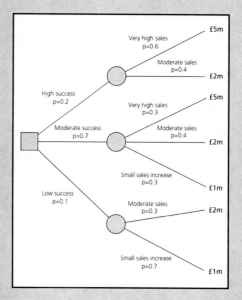

Working from the right to the left, for a highly successful campaign the expected return is:

£5m × 0.6 + £2m × 0.4 = £3.8m

Multiplying by the probability of a highly successful campaign (0.2) gives £0.76m.

For a moderately successful campaign the expected return is:

$$£5m × 0.3 + £2m × 0.4 + £1m × 0.3 = £2.6m$$

Multiplying by the probability of moderate success (0.7) gives £1.82m.

For low campaign success, the expected return is:

$$£2m × 0.3 + £1m × 0.7 = £1.3m$$

Multiplying by the probability of low campaign success (0.1) gives £0.13m.

Adding the three expectations together gives:

$$0.76 + 1.82 + 0.13 = £2.71m$$

Since this is below the £3m the campaign will cost, it is not worthwhile for the company to go ahead.

14 The three options can be set out as arms of the decision tree and each has three possible potential sales increases with their probabilities. Increases in profits vary depending on the ability of the factory to cope with potential increases.

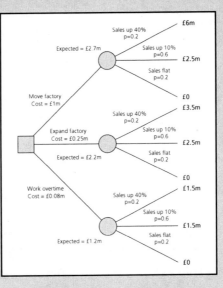

Multiplying each profits increase by the probabilities and adding gives the expected values shown on the diagram, for example, for 'move factory' we have:

$$£6m × 0.2 + £2.5m × 0.6 + £0 × 0.2 = £2.7m$$

For each option we can deduct the short term cost from the expected potential profits:

Move factory: £2.7m – £1m = **£1.7m**

Expand factory: £2.2m – £0.25m = **£1.95m**

Work overtime: £1.2m – £0.08m = **£1.12m**

Since the highest value is associated with expand the factory, this is the recommendation to the company.

15 Each of these terms is a number of combinations and can be calculated from the formula:

$$\binom{n}{r} = \frac{n!}{r!(n-r)}$$

$$\binom{4}{4} = \frac{4!}{4!0!} = \frac{4!}{4!} = 1$$

$$\binom{4}{3} = \frac{4!}{3!1!} = \frac{4!}{3!} = \frac{4.3.2.1}{3.2.1} = 4$$

$$\binom{4}{2} = \frac{4!}{2!2!} = \frac{4.3.2.1}{2.1.2.1} = 6$$

$$\binom{4}{1} = \frac{4!}{1!3!} = \frac{4.3.2.1}{3.2.1} = 4$$

$$\binom{4}{0} = \frac{4!}{0!4!} = \frac{4!}{4!} = 1$$

16 More combinations, but with larger numbers:

$$\binom{10}{6} = \frac{10!}{6!4!} = \frac{10 \times 9 \times 8 \times 7 \times 6!}{6! \times 4 \times 3 \times 2 \times 1} = \frac{10 \times 9 \times 8 \times 7}{4 \times 3 \times 2 \times 1} = 210$$

$$\binom{10}{1} = \frac{10!}{1!9!} = \frac{10 \times 9!}{9!} = 10$$

Note that we can cancel large amounts of factorials top and bottom, for example 6! in the first part, and 9! in the second part.

17 This is the number of different hands of playing cards that can be dealt out to four players:

$$\binom{52}{13} = \frac{52!}{13!39!} = \frac{52 \times 51 \times 50 \times 49 \times 48 \times 47 \times 46 \times 45 \times 44 \times 43 \times 42 \times 41 \times 40}{13 \times 12 \times 11 \times 10 \times 9 \times 8 \times 7 \times 6 \times 5 \times 4 \times 3 \times 2 \times 1}$$

$$= 635\ 013\ 559\ 600$$

18 Here we have a binomial situation, either a person bought a paper, or they did not. We have to assume that the probability is the same for each person (probably not true in reality).

a   P(4 of 5 bought a paper) =

$$\binom{5}{4}(0.25)^4(0.75)^1 = 5 \times 0.00390625 \times 0.75 = 0.0146484375, \text{ or about}$$
$$1.5 \text{ per cent}$$

b   P(1 of 5 bought a paper) =

$$\binom{5}{1}(0.25)^1(0.75)^4 = 5 \times 0.25 \times 0.31640625 = 0.3955078125, \text{ or about}$$
$$40 \text{ per cent}$$

c   P(no-one of 5 bought a paper) =

$$\binom{5}{0}(0.25)^0(0.75)^5 = 1 \times 1 \times 0.2373046875 = 0.2373046875, \text{ or about}$$
$$24 \text{ per cent}$$

19 Again this is binomial since the answer is either correct or incorrect. The probability of getting the correct answer by chance is 1 in 5, or 0.2. To find the required probability we have to find the chance of 4 from 10 correct:

$$\binom{10}{4}(0.2)^4(0.8)^6 = 210 \times 0.0016 \times 0.262144 = 0.088080384, \text{ or about}$$
$$8 \text{ per cent}$$

20 This may be treated as binomial since visitors buy or don't buy. The expected number of buyers is np, or $200 \times 0.005 = 1$. Finding the probability of no-one buying requires you to evaluate:

$$\binom{200}{0}(0.995)^{200} = 0.366957822$$

This is not recommended on a simple calculator, but easy enough on a spreadsheet.

21 Each of these answers can be found using the table in Appendix 2.

a   To the right of $z = +1.2$, we can just look up to get **0.1151**
b   Similarly for $z = +1.96$, we get **0.0250**
c   Here we have a negative z-score, so we initially find the area to the right of $z = +1.55$ which is 0.0606, and then subtract this from the total area of 1, we get **0.9394**
d   This is a similar question, so area to right of $z = +2.58$ is 0.00494, and the area to the right of $z = -2.58$ is $1 - 0.00494 = $ **0.99506**
e   Areas to the left of negative z-scores are the same as areas to the right of the same positive z-score, so here we can just look up the answer, which is **0.02275**

f   This is similar to part c, so we find the area to the right of $z = +1.5$ as 0.0668 and subtract from 1 to get **0.9332**

g   Firstly we look up the two areas: to the right of $z = +1$ is 0.1587, and to the right of $z = +2$ is 0.02275, we want the area between, so subtract the smaller one from the bigger one to get 0.13595

h   The areas outside these points will be the same, since the distribution is symmetrical, this is 0.025. Twice this is 0.05, and subtracting from 1 give **0.95**. This result means that 95 per cent of any normal distrubution is within 1.96 standard deviations of the mean.

22 To answer this question we need to convert the actual errors into z-scores using the formula:

$$\frac{\text{Value} - \text{Mean}}{\text{Standard Deviation}} = \frac{X - \mu}{\sigma}$$

a   Converting £125 gives:

$$(125 - 100)/20 = 1.25 = z$$

so the area above this is 0.1056 (from Appendix 2) which means that the probability of an invoice being over £125 in error is **0.1056** (roughly 10 per cent).

b   Converting £50 is:

$$(50 - 100)/20 = -2.5 = z$$

so the area below this is **0.00621**, which is the probability.

c   We start by converting both values to z-scores:

$$(75 - 100)/20 = -1.25$$

$$(125 - 100)/20 = +1.25$$

the area to the right of $z = 1.25$ is 0.1056
doubling this gives 0.2112
subtracting from 1 gives the answer of **0.7888**

WEB REFERENCE 👁

www.thomsonlearning.co.uk/businessandmanagement/curwin3

See the companion web site for further questions and annotated answers. There you will also find a PowerPoint presentation which you can use to help understand this area, or later, for revision. The site also contains links to other sites of interest.

Chapter 12

COLLECTING DATA

Why collect data? This is a good question when there is so much **data** about. We only need to browse through recent government publications to find lots of data. Most daily newspapers will have facts and figures somewhere. If we are working for a company, we only need to ask for recent sales, or prices or the number of days lost each month through sickness to begin to build a database (a collection of such facts and figures). It is true that we can find data on most topics of interest but the real issue is whether it answers the questions of interest to us. Essentially we collect data, if available sources fail to provide the specific details we are seeking. If we particularly want to know what certain people are thinking or doing we will need to ask them, ask about them or observe them.

OBJECTIVES

After reading this chapter and working through the exercises you should be able to:

- identify the important issues of data collection
- clarify the purpose of primary data
- describe the techniques of data collection
- critically consider questionnaire design

12.1 THE USE OF DATA

The data we work with will typically take the form of columns of numbers. We have seen how the calculations of a few descriptive numbers, *statistics*, can help us make sense of this. Your research may need the average waiting time for a particular service or most frequently requested size. Data is collected to inform the user; it then becomes **information**. Information is used in a variety of ways and is usually seen as an important element in the decision-making process. An important first step is to clarify the purpose of any research. It is likely that if we were asked to describe the facilities that a visitor could expect from a 'bed and breakfast' we would approach this differently from being asked to describe the recent experiences of visitors to 'bed and breakfast' accommodation. In the first case the focus of interest is the 'bed and breakfast' and we would want an adequate sample of such establishments. In the latter case, the focus of interest is the visitor and we would want a sample of such visitors. The focus of interest is called the **sampling unit** and we need to be sure that we are working with the right units. The unit of interest might be public service vehicles in the West Midlands or sports shops within a television region, or car owners in the UK.

12.2 THE FOCUS OF ENQUIRY

It is an important element of any enquiry to decide whether to use only the data collected by others or to add your own data. Data we collect is referred to as **primary data**. This data is collected to serve our specific purpose and may or may not have other value. It could be collected on specific customers, products or services, be commercially valuable or regarded as confidential. The data collected by others can come from a variety of sources and is referred to as **secondary data** (see Chapter 13).

We must clarify which people or which items are of interest to us. This will define the boundaries of our study. We cannot vaguely talk about children but need to state which children, whether these children are all of the same age or go to the same school or have the same ethnic background. Our research will only be accepted as meaningful if we have defined clearly those that we will work with. The group of interest is defined as the **population**. We need to be careful with this term. Typically, we are interested in what a distinctive group of people do or think, and this would be our population. However, it is not always people. The population of interest could be an estate of buildings, herds of animals or kinds of business. Having clarified those people or items of interest to us, we need to decide whether to obtain data from each one or just a selection. If we do decide to include all those of interest in our research, this is referred to as a **census**. A census is often thought of as the detailed enumeration of the population undertaken by the government in the UK every ten years, the most recent being April 2001. However, we use the term more broadly than that and include all sorts of distinctive groups: beer drinkers, a batch of car components, or trees in a forest. For non-governmental research a census is only viable if the population of interest is relatively small.

We may decide only to collect data from a sample of our population. Opinion polls, for example, will typically include between 1000 and 2000 respondents rather than all those entitled to vote. A quality control procedure may only require samples of 100 or 200 from batch sizes of several thousand items. A **sample** is a selection from the population of interest. There are a number of reasons why we might choose to select a sample rather than work with the complete population. We all need to work within time and budget constraints. Opinion polls tracking voting behaviour may need to be completed in one day, providing sufficient detail with sufficient reliability. A food manufacturer may want sufficient information to form a view on whether to test market a new type of packaging but not need precise statements from all possible consumers. The important point is that the data collection needs to be suitable for the purpose.

The numbers required for sample data depend not only on purpose, time and budget but also *variability* within the population. The 'more different' the population of interest, the larger the sample needs to be. If we are interested in the first landing place of a group of tourists boarding a plane (and can assume that no one jumps off!), we only need a sample of one. If we are interested in the 'first stop' of tourists entering an airport, or indeed the more complex data regarding final destination, then we will need a larger sample. The more *heterogeneous* (variable) this population, the larger the sample needs to be.

Having clarified the purpose of our research and the numbers required (sample size), we need to decide on the collection method. The deciding factor is likely to be the level of contact with the respondent. If the questionnaire needs to be covered in a professional and detailed way, our preferred method is likely to be personal interview. An interviewer will take time to identify the required respondent, perhaps making several recalls. A range of questions can be asked including those requiring open-ended responses and those of a sensitive nature. A trained interviewer can also prompt and probe to improve the quality of the answers. Given careful management, personal interviews can achieve a high level of response (perhaps between 50 per cent and 80 per cent depending on the survey). If self-completion is possible and relatively simple questions can be used, postal questionnaires have a number of advantages. Relative costs can be low and a good geographic coverage easily achieved. However, response rates can be low (perhaps around 30 per cent) and questions may be answered in a fairly superficial way. Telephone interviewing can be effective where personal contact is helpful and most respondents are willing and able to use a phone. Surveys of technical support staff and retail managers have successfully used telephone contact. If telephone contact becomes synonymous with 'cold calling' or other forms of selling then this will have a detrimental effect on the quality of response. Other possible methods include the completion of diaries, the use of email proforma and observation.

12.3 | SAMPLE SELECTION

If a population is relatively small and distinct, then sample selection is unlikely to be a problem. If a teacher is interested in the views of the lower sixth form on recent changes to the timetable, then the issues are relatively clear. The population (those students enrolled as lower sixth) is likely to be available in the form of a list. This listing is important and is referred to as a **sampling frame**. In this case the teacher will then need to decide on how many to select and how to select. The number selected will depend, as already discussed, on the time and resources (cost or budget are not always the best measures) available to the teacher, the range of views and the accuracy required. Suppose the teacher decides to select 50 from the enrolment of 500. There is still the question 'how to select'. Essentially, the teacher could look for some method of random selection or some form of purposeful sampling.

Randomness is an important concept and is used to justify much of what we say about our research. It should not be confused with a haphazard selection. A random selection gives each individual, or item, a *calculable chance* of being included. Once the rules of selection are accepted the method can be seen as transparent and fair. If the teacher were to select 50 names from the 500 placed in a hat by the lottery method, then this would be seen as fair. If, however, the teacher took the first 50 names on the list or those that attended the maths class this would be seen as potentially unfair or with **bias**, 'bias' being any systematic source of error.

It is through our method of sample selection that we can argue that our results are reasonable and fair. With acceptable results, we are able to present a generalized view, develop ideas and support decisions. You will be told about the importance of methodology (the methods used) in project work. Projects will often be justified in terms of their sample selection. Just to interview a few employers or people passing by in the street will generally be dismissed as lacking validity and, at best, give misleading results.

12.4 | SAMPLE DESIGN

An important element of sample selection is *sample design*. In some cases, a simple random selection will do, and in others cases we will look to improve our approach by a more complex design. Sometimes retaining a random method of selection is of overwhelming importance and in other cases we will accept that an interviewer can choose 'typical' people.

Essentially, sample selection must be appropriate for the purpose of the survey. We will try to retain elements of randomness in all design and try to improve on the basic lottery method.

Simple random sampling (SRS) gives all members of the population an equal chance of selection. Remember that a population is not necessarily all the people but a distinct group of people or items of interest to us. Our survey may only be concerned with the manual workers employed by a company or those students in lower sixth taking IT. A listing of those of interest is called a *sampling frame*. A simple random sample could be achieved by picking names from a hat. However, we are more likely to work with a list where each entry has been numbered from one onwards. If we consider a university with 10 000 students, for example, we are likely to work with some method of random number generation rather than look for a big hat!

We can use computer generated random numbers or random number tables. A random number table could take the form shown in table 12.1.

87 024	74 221	69 721	44 518	58 804
04 860	18 127	16 855	61 558	15 430
04 852	03 436	72 753	99 836	37 513

TABLE 12.1   An extract from a random number table giving random digits from 0 to 9 arranged in blocks of 5 for convenience

If we need to select from 10 000 students we would number them 0000 to 9999 and take blocks of 4 digits from our random number table. The first student selected would be numbered 8702 and the next one 4742, and so on. In practice we would not always start from the beginning of the table, or take the same sequence from a random number generator as this would not be random. In the case of a table we could simply cross out used numbers (see table 12.2).

~~87 024~~	~~74 221~~	69 721	44 518	58 804
04 860	18 127	16 855	61 558	15 430
04 852	03 436	72 753	99 836	37 513

TABLE 12.2   An extract from a random number table showing the deletion of 'used' numbers

Stratified sampling attempts to use our knowledge of the population to improve the results of our survey. Relevant groups or *strata* are identified before sampling begins, and samples selected from within each of these strata. Suppose, for example, we were interested in attitudes to smoking in public places, then we might want our sample to reflect the proportion of smokers and non-smokers in the population. If we were interested in transport policy issues, we might want a fair representation of people using various modes of transport. The selection of people within a strata can be random using the lottery method or some method of random numbers.

Suppose we wanted to select 100 students from a university of 10 000 students. Suppose also we had the additional knowledge that the university had three faculties with 4000 students in the business school, 4000 students in humanities and 2000 students in engineering. It is also thought that in terms of our research the experience of students could differ substantially between faculties.

In this case, we could argue that the relevant strata were faculties and select 40 students randomly from the business school, 40 students randomly from the Humanities and 20 students randomly from engineering.

Some populations will have natural groups or *clusters* that we can use for sampling purposes. We may also cluster sample for practical reasons like ensuring that interviewers can be reasonably allocated to those selected. It is easy to image the complexity of having selected individuals spread across the UK if personal interviews were seen as essential. If however, we had decided on a postal method of enquiry, then geographic spread would be of no importance. In some situations we might choose to select families rather than individuals. If we were interested in the diversity of fish being taken from the North Sea, we might first select boats and then examine all of the catch. A *random walk* is an interesting variant of cluster sampling. An interviewer is given one or more starting addresses and asked to follow a procedure, selecting every fifth house thereafter.

Systematic sampling is seen as another variant of cluster sampling (a linking of selected individuals, items or households). Suppose we wanted to select 40 students from a business school with 4000 students. Effectively we would want to select 1 in every 100. If these students were in a numbered list from 0000 to 3999, we would make a random start from 00 to 99 (00 and 0000 are still zero) and take every hundredth thereafter. If the next two random digits from our random number table or generator gave 21, we would select those numbered 21, 121, 221, 321 and so on. This method can work effectively provided there is no systematic pattern within the sample frame. You could image a situation were a list had been constructed in terms of Mr and Mrs, and the sampling interval selected all male or all female respondents (an example of systematic bias).

For large complex samples, we typically choose a multistage design. The sampling process takes place in stages. At the first stage we could select some or all of the regions. These regions could correspond to television regions or other administrative boundaries. Parliamentary constituencies are commonly used at the next stage. The number selected will depend on the size of the sample required but will generally be four or more. Some stratification might take place at this stage to ensure a representative spread of selected constituencies. At the next stage, typically households or individuals are selected from the electoral register. It is likely that some variant of systematic sampling will be used to select from a list as lengthy and detailed as the electoral register.

In designing a sample, we are likely to seek improvements on simple random sampling for practical reasons or as an attempt to make the sample more representative. Retaining the element of randomness will ensure that the proposed design is accepted as valid and without bias.

However, for a variety of reasons, we may abandon the requirement that each individual, or item has a calculable chance of being included. If, for example, a market research company needed to talk to recent users of a particular product or people who own a particular kind of pet they would have no convenient sampling frame to work with. It could be prohibitive in terms of time or cost to produce such a listing. The choice might be to do the research without random selection from a sampling frame or not do the research. Even if random selection is not used at the final stage, a number of constraints are likely to be imposed to ensure that the sample is more representative.

The most usual form of non-random sampling is **quota sampling**. Important characteristics are identified and interviewers are required to select respondents in given categories. Typically, respondents sought would be specified in terms of age, gender and occupation. Categories for respondents could also include those of interest to the survey, such as smoker/non-smoker or car owner/non-car owner. In general we would want the achieved sample to reflect the proportions in the population. There are clearly reasons to be concerned about quota sampling – the interviewer will have the final choice and may introduce bias, for example to avoid men with beards. However, quota sampling does seem to work in practice for many kinds of research, particularly consumer market research.

12.5 | SAMPLE SIZE

The size of the sample required will depend on a number of factors, including:

- the accuracy required
- the level of difference within the population
- the detail required, particularly from subgroups within the population

If, for example, an accuracy of ± 3 per cent is required rather than ± 5 per cent then a larger sample will be needed. This is why it is important to clarify the purpose of the survey and the requirements of the end user. If we need to use the results of the survey to make critical inferences about a much larger population, then the size of the sample and the design will become more important. The sample size requirements would increase considerably if we wanted to say something about UK students in general, rather than the students at one particular business school.

The more diverse respondent characteristics and views, the more important it becomes to look for improvements in sample design and increase in sample size. If we decided to define our population of interest as UK full-time and part-time students, including students in further and higher education rather than just UK students in higher education, then again we would need to consider a more complex design, possibly multi-stage, and a larger sample size.

If we also wanted to look at certain sub-groups in particular, for example mature students or international students, then again we would need to ensure that numbers were sufficient.

12.6 | ASKING QUESTIONS

Information will come from the questions we ask. The general advice is that good questions are far harder to write than you might think. The order of questions, the content, the structure and the wording can all make a difference to the usefulness of the answer. A *questionnaire* is the tool used to work with respondents, take them through a flow of questions and collect a few necessary personal details. The questionnaire will involve a potential respondent in our research and provide the information like contact address to check validity, if required. It is regarded as good practice in many market research organizations to check with a small proportion of respondents that the interview took place and that responses were accurately recorded. However, the assurance of confidentiality still needs to be retained. The term 'questionnaire' has been used in a variety of ways. In some cases it has been taken to mean exclusively a survey where the respondents fill in the answers, for example a postal question. We use the term questionnaire more generally to mean all data collection techniques where respondents are asked to respond to the same set of questions. Questionnaires will therefore provide the basis of data collection for email, postal and telephone surveys and for structured and semi-structured interviews. A special case is likely to be unstructured interviews where the listing of questions is usually referred to as a *checklist*. The interviewer will cover all the points with each respondent but will be looking for more detailed answers that may be recorded for future analysis. The interviewer may also discuss the answers and elaborate on the questions.

The choice of questionnaire will depend on a number of factors including:

- purpose of the survey (which should be clarified by objectives)
- availability of respondents
- likely response to proposed questions
- types of question necessary to generate required data
- amount of detail required

Suppose, for example, we were interested in the experiences of new students at a particular university. We would need to clarify what we meant by 'experiences', which is likely to include course information provided, accommodation, course induction, course timetable, contact with the course team and teaching. The potential respondents are likely to be interested in the purpose of the survey and are likely to be articulate. If the university is supportive of the survey, students could be contacted by post, by email and by going to the classes. Using classes as a means of contact could introduce a source of bias. It could be the case that those students unhappy with their time at university may have stopped attending or attend infrequently. A range of question types could be used but the chances are that the respondents would be given the opportunity to elaborate on their answers rather than just being allowed a yes/no response. A mix of questions would provide both classification type data (age, sex, course, previous qualifications) and comment ('In what ways ...', 'In your opinion, what ...'). It is also worth remembering that a survey may complement other methods of research so, for example, we may carry out a survey with a 100 first-year students and in-depth interviews with a further five. The numbers would clearly depend on the purpose the research, resources available and information required.

The design of the questionnaire will depend on the level of contact as shown in figure 12.1.

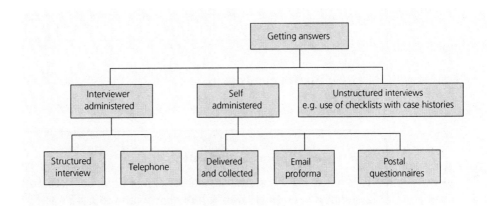

FIGURE 12.1   Uses of questionnaires

The questions developed for use with a postal questionnaire are likely to have very different characteristics from those used in a structured interview. In a postal questionnaire we would be mindful of the time respondents are willing to give and the importance of question clarity. A postal questionnaire would be seen as a cost effective way of undertaking research where questions would mostly give a choice of alternatives. In contrast, we would probably justify the additional cost of structured interviews by the need to get more in-depth answers over a broader range of questions and the ability to obtain observational data. A structured questionnaire would also allow a trained interviewer to record the answers to more open questions.

When producing questions it is useful to adopt good practice from elsewhere. It might be possible to adapt questions from other questionnaires, for example. In most studies it is possible to find a questionnaire that has been used before and attempt to modify and improve it.

Possible types of question are shown in figure 12.2.

An *open-ended question* allows respondents to answer freely in their own words. The interviewer will be expected to record what was said verbatim. These questions are very good at exploring ideas but can be difficult to analyze. Some respondents talk at length and it can be difficult to record everything they say.

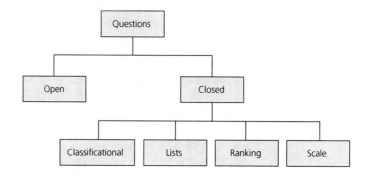

FIGURE 12.2   Types of questions

Examples of open-ended questions:

1. What were the main reasons for choosing to come to university?
2. What do you particularly like about your course?
3. In what ways has the organization changed in the last 12 months?

Closed questions will give the respondent a choice of predetermined answers but may add 'other please specify …'. These questions are typically quick and easy to complete and easy to code for computer-based analysis.

Questions can be used merely to *classify* respondents. The inquirer may be interested in gender, age, marital status and occupation. The questionnaire may also want to establish whether or not respondents have particular characteristics.

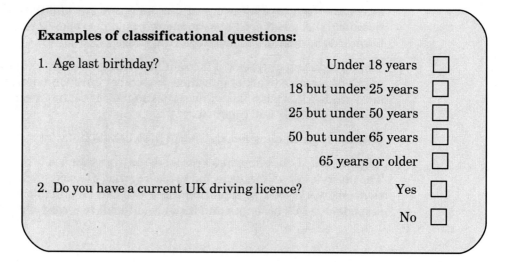

List questions ask the respondent to select one or more from a given list of alternatives.

Examples of list questions:

1. Please specify your main mode of transport to and from work.

   Car ☐

   Train ☐

   Bus ☐

   On foot ☐

2. Do you have a current UK driving licence?   Yes ☐

   Other (please specify) ....

This question only allows for one tick as it does ask for main mode of transport. You could extend the list by adding other modes of transport (e.g. bicycle) or being more detailed (e.g. type of train).

Which are your favourite activities on holiday?

Going to the beach ☐

Reading ☐

Sitting around the pool ☐

Sightseeing ☐

Sunbathing ☐

Swimming ☐

Going for walks/walking ☐

Watersports ☐

This question allows one or more ticks as a respondent may have a number of favourite activities. Again the list can be extended to include a wider range of activities and could also include 'Other (please specify)'. It can be difficult to decide what to include in such a listing. We could adapt a listing used in a previous survey, which may have the advantage of compatible data, or we could do some preliminary research, perhaps with small groups of respondents, to identify common interests.

A *ranking question* will ask a respondent to give an order or preference. These questions are particularly helpful if you want to establish relative importance.

Examples of ranking questions:

1. Rank the following factors in order of importance when deciding to buy (from 1 being most important to 5 being least important)

Rank

Reputation of company

Printing quality

Price

Delivery

Range of options

A *scale* or *rating question* is often used to determine the strength of views or opinions.

Examples of scale questions:

1. How likely are you to try this product?

Very likely	1
Quite likely	2
Neither likely nor unlikely	3
Quite unlikely	4
Very unlikely	5

Respondents could be asked whether they agree or disagree with particular statements:

	Agree strongly	Agree slightly	Neither agree nor disagree	Disagree slightly	Disagree strongly
This pack looks expensive	1	2	3	4	5
This pack looks attractive	1	2	3	4	5
This pack looks practical	1	2	3	4	5

There are many ways to structure questions, but we hope the few examples given will be a useful guide.

12.7   FRAMING QUESTIONS

The quality of answer, and ultimately the data generated, will depend on the question wording. A question will need to be understood by all possible respondents, have a single interpretation and not offend.

When wording questions try the following guidelines:

- *Avoid bias*. Keep the questions neutral and avoid leading the respondent in a particular direction. Questions like:

    'Would you agree that ...'

    or:

    'Given the excellent product range ...'

- *Avoid jargon or shorthand*. Questions need to be written in a form acceptable to respondents. Questions can easily be confusing:

    'Would a reduction in APR ...'

    or:

    'What would you see as the advantages of a buddy system?'

- *Avoid ambiguous words*. Words like 'usually', 'frequently' or 'recently' can all be interpreted differently. Try to be more explicit by asking:

    'How often in the last week ...'

- *Keep questions short and simple*. Long questions can lose the attention of the respondent and often create a more complex scenario. A series of shorter questions is generally seen as more effective.

    How would you respond to a question like:

    'Given the recent change in price and the fall in demand usually experienced at this time of year, do you think a 20 pence promotional voucher would ...'

- *Avoid questions with negatives*. Questions like:

    'Do you ever ...'

    are seen as more effective than

    'Do you never ...'

- *Avoid hypothetical questions*. Generally respondents can only give reliable answers to what they know. We generally discount questions like:

    'If you were to earn an extra £20 000 a year, would you ...'

- *Use filters to avoid irrelevant questions*. If a set of questions only applied to non-smokers you could add:

  'If a smoker jump to question …'

- *Make it easy for respondents to answer the questions*. Questions like:

  'Have you bought any of the following items in the last week'

  are easier than:

  'What did you buy in the last week'

  Giving a range can be more helpful than asking for a specific value. We could ask:

  ' … did you spend under £10 / between £10 and under £20 …'

  rather than:

  'How much did you spend …'

In general it is easier to use or improve existing questions than write them from scratch.

12.8 | SOURCES OF ERROR

Having collected data we can never be completely sure that we have got the right answers. There are essentially two types of error that we need to consider and minimize if possible: **sampling error** and **non-sampling error**.

If we take a second or a third sample the chances are that we will get slightly different results. By definition, a sample will not include everyone, and the answers will depend on those selected. Image a carefully designed survey of 1000 possible voters. The design and the sample size would lead us to expect reasonable results. However, each time a sample is selected it can contain a different mix of respondents. If the method works then we would get the same kinds of results. If we are working with random samples then theory does allow us to say how good our results are (too advanced for this book, but covered in Jon Curwin and Roger Slater, *Quantitative Methods for Business Decisions*, 5th ed., Thomson Learning, 2002). This variation from sample to sample is known as *sampling error*, and good sample design and a reasonable sample size will allow us to keep this small.

There are a number of possible errors than cannot be explained by the sampling process and these are referred to as *non-sampling errors*. Selected respondents can refuse to participate. Non-response can be a major problem. Some surveys report response rates as low as 30 per cent. The main problem is we don't know if the non-respondents differ in any significant way to the respondents. Improved training for interviewers, improved documentation and incentives are all seen as ways of improving the response rate. You will only get answers to the questions you ask and the answers will depend on how you word the questions. A recent survey looked at household finance and suggested links between the number of young adults staying at the parental home and their level of income. Others have suggested that such a link could not really be made unless other factors were also considered, such as the more liberal attitudes in the more affluent parental homes. One critic said 'it's more about sex than finance but they found what they were looking for'. Another source of error that we should have some control over is *transcription* error. If responses are incorrectly recorded or copied then this will affect the quality of the data. Checking, cross-referencing and common sense should reduce this form of error. However, examples have been seen of drivers aged 6 months and £112!

12.9 CONCLUSIONS

It is likely that we will find some data on most topics of interest, and this we call secondary data. However, it is unlikely that this will provide all the information that we are looking for all of the time. The data we collect is referred to as primary data. As soon as we want more detailed or more recent information on usage, attitudes and opinions we need to consider ways of collecting it. In some cases it may just be a matter of observation, like the number of people waiting for service or the number of people not wearing car seat belts. In most cases however, we need to develop a number of questions and need to find ways of presenting these to selected individuals. These individuals can be randomly selected in some way or can be chosen with some discretion by an interviewer. We try to work with random methods of selection where possible or retain some elements of randomness. To justify our results we need to argue that our sample was reasonable and representative. We also need to ensure that the questions asked are meaningful to the respondents and enable a completeness and correctness of answer.

When you have worked through and understood the chapter you will be in a position to:

- understand the importance of collecting primary data, particularly in a business context
- understand a number of the techniques used for the collection of data and discuss their application
- know the basic design features of a questionnaire
- work to improve the wording of questions used in typical questionnaires

1   Give examples of why primary data might need to be collected.

2   What would be the sampling unit of interest for research into:

- MOT failure
- the facilities available for the disabled at private fitness clubs
- the use of bicycles for journeys to work

3   Give reasons why you might use a survey as the preferred method of enquiry for the following topics for research:

- the public response to the siting of a new airport
- the out of school activities of children aged 11 to 13 years
- the level of support offered to the young homeless

4   What factors would you need to consider when deciding the sample size for a survey on the impact of student fees?

5   How would you decide whether to use personal interview, postal questionnaire or telephone interview for a sample of:

- new car owners
- users of internet banking services
- dog owners
- health care requirements of the elderly

6   How would you develop a sampling frame for fast food outlets within walking distance of a local school?

7   Comment on methods of sample selection for the following:

- to interview a cross-section of motorists about the advantages of diesel engined cars
- to give out at least 200 self-completion questionnaires to customers of a local swimming baths regarding opening times
- to phone a range of local businesses about job opportunities for school leavers

8   What method of sample selection would you recommend for the following:

- recent school leavers
- purchasing managers in an IT environment
- those aged 25 or under that regularly attend professional football matches

9   How would you decide on the sample size for:

- a survey of lower sixth-form students within a particular school regarding the quality of catering facilities
- a survey of privately owned vehicles regarding the condition of tyres
- a survey of individuals involved in a minority sport

10 Draft a few questions to find out:

- how much time students in your class spend on study
- where students in your class have taken holidays
- what concerns other students have about personal security

11 Improve the following questions:

Q.1. Do you consider your lifestyle healthy?   Yes/No

Q.2. Do you regularly buy fruit and veg?   Yes/No

Q.3. Do you go jogging or take othe forms of exercise?   Yes/No

If yes, what kinds ....

Q.5. What age are you? ....

Q.6. How much television do you watch in your leisure time?
                                                        .... Hours

12 Describe how you would attempt to reduce both sampling and non-sampling error in a survey of those people that regularly use a local gym.

✔ 12.11 ANNOTATED ANSWERS

1 The main reason for collecting our own data is that existing data is not sufficient for our purposes. It may be that existing data is not specific enough for a product or service of interest to us. Market research is mostly concerned with providing up-to-date business information to companies on products and services of interest to them. Data will also age, and we may wish either to validate existing data or add more recent figures.

2 Sampling units of interest:

- If our enquiry were concerned with MOT failure, the sampling unit of interest would be suitably defined vehicles that had recently failed the test. The sample design would be concerned with how these vehicles were selected.
- If our enquiry were concerned with the facilities available for the disabled at private fitness clubs, the sampling unit of interest would be suitably defined clubs. We would then audit selected clubs in some way to assess facilities.
- If our enquiry were concerned with the use of bicycles for journeys to work, we would need to decide whether we were just concerned with those that are currently users, all potential users, or all those that make journeys to work. As always, the clarification of purpose is important. We need to decide whether we want to ask only those that

have current experience of using bicycles for work, or wish to find out also why some people would never consider using a bicycle for their journey to work.

3   The reasons why a survey might be the preferred method of enquiry for the following topics for research include:

- For the public response to the siting of a new airport – consider responses to a range of issues that impact on such a decision. A survey should clearly identify the 'population' or 'populations' of interest. Surveys could be repeated to monitor whether views are changing over time. Survey methodology can present the views of those that perhaps do not speak up at public meetings.
- Out of school activities of children aged 11 to 13 years – it is likely that this information is not kept elsewhere. Schools may have records of activities that take place during school time and may have some knowledge of some of the activities that take place outside school time. However, they may see these as confidential and may not make them more widely available. It is also difficult to assess interest in minority sports and other interests. You also need to consider whether it would be more appropriate to interview a parent or how parental consent could be obtained.
- The level of support offered to the young homeless – you should be able to find helpful secondary data. A survey should be able to add content to existing data and give more recent information. A survey can also investigate an aspect in more depth, like a change in the law or the impact of a local initiative.

4   To decide the sample size for a survey on the impact of student fees, you would need to clarify whether you were interested in all possible students or a particular group of students. You would also need to clarify whether coverage was restricted to one institution, a type of institution (i.e. universities) or if you need to achieve a national coverage. The more ambitious you survey the larger the sample required. You would need to consider the required accuracy. If you wanted to make very precise statements about the general population of students, then again you would need to consider increased sample size. The size of the sample would also depend on how views amongst students varied. The more varied the range of the views, the larger the sample required to reflect this. As a guide, for a national survey of this kind, you should be thinking about a sample of 1000 plus that is stratified to ensure better representation.

5   The likely reasons that would inform the decision of whether to use personal interviews, postal questionnaires or telephone interviews for the following include:

- For new car owners – a willingness to respond (new car owners like to talk about their new cars!), in most cases the questions can be easily structured and may only require ticking and the people of interest might be geographically spread. If a suitable sampling frame exists, then postal questionnaires are often very cost effective and give reasonable rates of response. If a sampling frame is not available, then we could use some form of quota sampling.
- Users of internet banking services – if a sampling frame of email addresses were available we would probably send an email questionnaire (a variant of a traditional postal questionnaire). Again, if a sampling frame is not available, then we could use some form of quota sampling.
- Dog owners – unless we do the obvious thing and take a quota sample of those people seen with dogs (which could be very cost and time effective) we would need a 'filter question' to determine dog ownership. It is likely that we would look for alternatives to the more expensive personal interview. We would need to be cautious of any survey that only included those that walked their dog. This approach could exclude a number of dog owners particularly those that own a dog for company or because of limited mobility.
- Health care requirements of the elderly - we would first need to decide who to interview. Should we interview elderly persons or should we interview informed health care workers? In research of this kind we often turn to the expert witness. If we decided to interview a representative group of the elderly, it is likely that we would choose the more expensive method of personal interview because of the difficulty possible respondents might have with either postal questionnaires or telephone interviews.

6　Given that the number of fast food outlets within walking distance of a local school is likely to be relatively small and what may be regarded as walking distance will also depend on possible routes, we would need to combine a knowledge of acceptable distances (perhaps from a survey of school pupils) and a mapping exercise. A suitable listing would provide our sampling frame. We might wish to add descriptors, such as 'fish and chip shop', to enable stratification to take place if required.

7　Comments on the following methods of sample selection could include:

- To interview a cross-section of motorists about the advantages of diesel engined cars – not all motorists would be interested in diesel engined cars and not all motorists would claim to know about the advantages of diesel engined cars. In most surveys we would also aim for some kind of balance and also be interested in the perceived disadvantages.
- To give out at least 200 self-completion questionnaires to customers of a local swimming baths regarding opening times – it all depends on

whether these 200 questionnaires reach a representative sample. It could be that the questionnaires are taken by a particular group of customers (i.e. the first 200 to arrive on the day of issue) and these do not reflect the range of customers throughout the week. The survey would also exclude those that do not use the swimming baths because of the opening times.

- To phone a range of local businesses about job opportunities for school leavers – telephone methods are known to work effectively with businesses and may provide useful information. We would need to clarify whether we were just interested in job opportunities from local businesses or whether we were interested in other job opportunities such as not-for-profit organizations or businesses that were not considered to be local.

8 What method of sample selection would you recommend for the following:

- Recent school leavers – this will depend on the scope of the study. If we were only interested in those that had left a particular school in the last year, we could use the available records (as our sampling frame) and send a postal questionnaire. The chances are that the addresses would be sufficiently accurate and there would be sufficient interest to make this approach worthwhile. If our interest extended beyond one school (a wider scope), then we would have to consider all the design issues of a larger survey.
- Purchasing managers in an IT environment – because of the nature of the profession and the variety of the workplace this would be a difficult population to define and the construction of a sampling frame would present many operational issues. We could first select organizations and then select purchasing managers from those organizations. Given the type of respondent, we would probably try to develop an electronic (email) approach.
- Those aged 25 or under that regularly attend professional football matches – given the age and interest constraints, and probable lack of a sampling frame, we would look to develop some form of quota sampling.

9 How would you decide on the sample size for:

- A survey of lower sixth form students within a particular school regarding the quality of catering facilities – given the clearly defined group and the small numbers involved we would be tempted to include them all. In general, we would be looking for a minimum of about 50 representative of the range of students.
- A survey of privately owned vehicles regarding the condition of tyres – the sample size would clearly depend on the purpose of the survey. If we were trying to say something about vehicles in general across

the UK, we would probably be looking at a stratified survey of at least 1000 vehicles.

- A survey of individuals involved in a minority sport – this would depend on the numbers involved. The availability of a sampling frame would be important. If we were looking at just a few individuals (e.g. bog snorkelling) then we would be tempted to include them all. If however, we were looking at several hundred or several thousand, the chances are we would either want a reasonable proportion or at least a couple of hundred.

The technical calculations are not covered in this book but can be found in Jon Curwin and Roger Slater, *Quantitative Methods for Business Decisions*, 5th ed. (Thomson Learning, 2002).

10 Drafting questions

It's only when you come to write questions do you realize how difficult it is. It is also easier to improve an existing questionnaire than writing questions from scratch. When drafting questions you need to be sure that they will be understood by the respondents, sufficiently precise and not too long. It is unlikely that questions will be right first time and you should look at an approach that allows for question testing and improvement. It is seen as good practice to trial a questionnaire on a relatively small sample, often about 30 respondents. This testing of the questionnaire and other aspects of the survey (e.g. the adequacy of interviewer briefing) is known as a *pilot survey*.

This question was designed to give you a chance to try writing a few questions and to experience some of the difficulties.

11 Possible improvements to the following questions:

Q.1. Do you consider your lifestyle healthy?   Yes/No

The response of yes or no is not very informative and ignores many of the lifestyle issues. Some may consider their overall lifestyle healthy but smoke while others would consider this incompatible. You could ask a number of 'build-up' questions like 'Do you participate most weeks in a sport', or you could ask the respondent 'Which of the following statements best describes you' and then give a range of statements.

Q.2. Do you regularly buy fruit and veg?   Yes/No

To start with this is *two* questions – they may regularly buy one but not the other. You would also need to consider the difference between buying and eating. The term 'regularly' is rather vague and could be subject to differing interpretations. If you were interested in personal consumption you could ask questions like 'In the last seven days, which of the following have you eaten' give a list, and for those items included ask a 'follow up' question on 'How often ...'.

Q.3. Do you go jogging or take othe forms of exercise? Yes/No

If yes, say what kinds ....

It is seen as good practice to ensure the correctness of question structure, grammar and spelling (the r is missing from 'other'). In this question, the respondent could be 'led' by the early reference to jogging. You could ask whether the respondent took any form of exercise, and 'if yes' ask what exercise do you take and perhaps how many times a week.

Q.5. What age are you? ....

It is more usual to ask for 'Age last birthday' or to give the respondent ranges i.e. '18 but under 21 years'. Some respondents are reluctant to give personal details and you do need to consider carefully when and how you ask such questions.

Q.6. How much television do you watch in your leisure time? ....Hours

With this kind of question you need to consider what and when, and whether a term like 'leisure time' is sufficiently clear. You also need to remember that on a number of issues recall may be poor. Can you remember what you watched on television three days ago?

12 Describe how you would attempt to reduce both sampling and non-sampling error in a survey of those people that regularly use a local gym.

To reduce sampling error can have cost and time implications. Essentially, sampling error can be reduced if we can increase sample size and improve our use of stratification factors. To reduce non-sampling errors we would need to consider the way we selected and approached possible respondents, attempt to improve question wording and consider issues like interviewer training. The key elements include ensuring some randomness, try to keep the sample representative and not to 'lead' the respondents in the type of reply they give.

WEB REFERENCE ◉
www.thomsonlearning.co.uk/businessandmanagement/curwin3

See the companion web site for further questions and annotated answers. There you will also find a PowerPoint presentation which you can use to help understand this area, or later, for revision. The site also contains links to other sites of interest.

Chapter 13

USING OTHER SOURCES OF INFORMATION

Data exists in many different forms. It may be the outcome of a simple recording system (the number of absent employees on a particular day), or the result of a dedicated piece of research (the incidence of cancer as it relates to the measurement of other factors that might be predictive). We make a useful distinction between data and information. *Data* is a collected set of facts or opinions, often in numerical form. *Information* is data organized in such a way as to inform the user. What is important is that data becomes informative. When using other sources of information we are not just looking to add to data but looking to add to the content of our work. Other sources of information are particularly good at developing the historical perspective and providing general background information. This puts our work into context, knowing the work of others (and being able to reference the work of others).

Collecting our own data does have advantages (as seen in Chapter 12). It will relate specifically to our problem of interest and should be up to date. This data is referred to as *primary data*. One major limitation is often the range of coverage achieved. Data collected by others might be sufficient for our purposes. Data from other sources, including government and professional market research agencies, might also be informative enough that we reconsider the focus of our own enquiry and reconsider the data we need. This is called *secondary data*.

OBJECTIVES

After reading this chapter and working through the exercises you should be able to:

- identify useful sources of secondary data
- manage the process of searching for data
- appreciate the availability of information from the web
- value the contribution of additional information

13.1 | BROWSING

There can be an irony in the research process. Research can be described simply in terms of a series of steps including the setting of objectives and the collection of primary and secondary data. However, the very process of doing research should make us better informed. We should be learning by doing. As we become better informed, we might wish to revisit the objectives of our research and the data needs. We may even wish to revisit the project title if some more valuable or insightful area of research emerges. It can often be the case in real research that we need some knowledge of the 'facts and figures' to develop an awareness of the data we really need.

There is also a good chance that whatever the research topic, there is useful work out there waiting to be found. Even if there is a clear case for using a questionnaire with our population of interest, it is likely that there are results from previous surveys that we can use for purposes of comparison. Previous research may also help us to define more clearly the things we want to measure. Once we have a knowledge of previous surveys, there is little point in trying to invent a new questionnaire when many aspects of structure and working have been tried and tested. Experience suggests that it is easier to improve the wording of existing questions than write questions from 'scratch'. To *browse* is seen as a legitimate research technique. Browsing is also seen as a good way of addressing the problem of not knowing what to do next. Have a good look around – you never know what you might find!

Good places to browse include:

- newspapers, magazines and journals
- book shops
- published statistics, say in a library
- companies' annual reports and accounts
- the records kept by organizations for internal purposes
- electronic databases
- the internet

Recently, a student group had been trying to develop ideas about the impact of change on the United Kingdom. They had managed to collect volumes of material from the internet and still needed to decide how to collate a disparate mix of data. Whilst looking for a recommended book in the business section of the local bookshop, one member of the student group came across a book by Richard Scase, called *Britain in 2010 – The New Business Landscape* (Capstone, ISBN 1-84112-100-2). Not only did this book use a range of data effectively but also used, in context, methods of analysis like linear regression (see Chapter 8). This book was there waiting to be helpful to this student group but needed to be found. Finding the right secondary sources of information can be a matter of luck but you can improve your luck by 'keeping your eyes open'.

13.2 | MAKING A START

There is an old saying that the first step is the most difficult. We offer you the following advice based on experience rather than scientific research:

- *Make any start* – The chances are that once you do start collecting secondary data you will quickly become more informed and more selective. Data search is not only about finding data but also knowing what data to reject. As the nature of the research topic becomes more clearly defined and research objective refined, criteria will begin to emerge that will confirm inclusion or give reason for exclusion.
- *Look at the statistics published by government* – An immense range of statistical information is provided by government (see **http://www.statistics. gov.uk**) covering topics as diverse are population, health and the environment.
- *Look at other sources of information* – Professional associations, political parties, trade unions and charities are all possible sources of information. Think about the nature of your research and ask the question 'Who would have an interest in this?'
- *Use the internet (selectively)* – What characterizes the internet is the volume of output. Some sites are particularly good (try **www.bbc.co.uk** for example) but you must question the validity of those that you are not sure about. It is worth knowing who owns the site and what their purpose is.
- *Use your library* – Not only will a library provide valuable source information but you should also get helpful advice. You could also check what training the library offers.
- *Look through the books in your local bookshop* – There is always a good chance that you will find something useful as you browse through the new books and latest editions. However, 'Curwin's rule' might come into play – the book you most want is also the most expensive! You can of course go online and look at **http://www.amazon.co.uk**.

As you look, you will begin to find more and more data. Too much information is not a problem to begin with. However, it can become a problem and you will need to become more and more selective.

Brainstorming can be useful at any stage of research. You may need to brainstorm ideas to get a good project title to work on. You may need to brainstorm to develop your ideas of how to proceed. Brainstorming can refer to a range of approaches including a creative pause when you want new or even wacky ideas to emerge to a professionally managed idea generation workshop.

One student group asked to develop ideas about the impact of change on the United Kingdom used brainstorming to come up with the following ideas:

Ask at the library   *Try the internet*   *Talk to a lecturer*

Read some science fiction   *Find a book on managing change*

Look for a TV programme   *Look at The Economist*   *Ask a politician*

Get information from the political parties   *Find out plans of business*

Talk to Greenpeace   *Talk to anyone*   *Talk to a cross-section*

Find out what young people want   *Find out what is happening in America*

Why follow America?* * Look at the health of people*

Take the age now and just add on some years   *Identify the new technologies*

Low manufacturing costs in developing world   *Growth of developing countries*

The euro

* one principle of brainstorming is to defer judgement. If we try to evaluate during brainstorming we could inhibit the flow of ideas. The aim is to capture an idea and then move on to more ideas. Evaluation can always be done later.

We would expect a good brainstorming session to give 30 or more ideas.

To bring some kind of structure to your data search you could try using a *relevance tree*. They have a hierarchical type of structure and can look like an organisational chart. You start with a general statement, such as the UK in the year 2010 and begin to work downwards. Figure 13.1 shows how we might start to develop a relevance tree.

A relevance tree is a good way to bring ideas together and can give a structure to any report that we may plan to write.

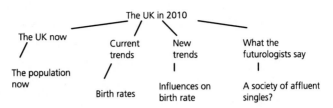

FIGURE 13.1   The beginnings of a relevance tree

13.3 | USING OFFICIAL STATISTICS

A good place to start is to look at the statistics produced by government. Try the Office of National Statistics (ONS) web site (**http//www.statistics.gov.uk** – see figure 13.2).

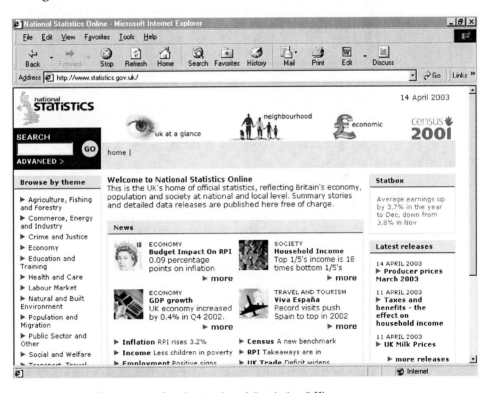

FIGURE 13.2   The front page for the National Statistics Office

Suppose you are interested in the average number of children in a family. A typical search might include looking for 'Social Trends' (using the search box provided), then looking at the 'Social Trends 30 (Datasets)', then 'Living in Britain' and eventually getting to an Excel spreadsheet (figure 13.3).

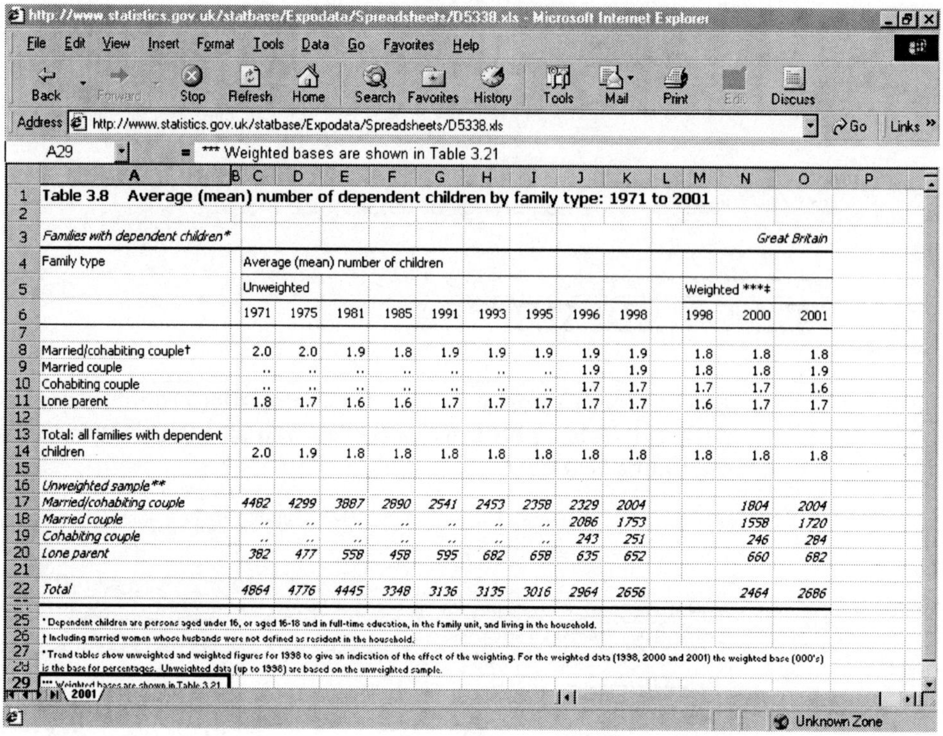

Table 3.8 Average (mean) number of dependent children by family type: 1971 to 2001

*Families with dependent children** — *Great Britain*

Family type	Average (mean) number of children											
	Unweighted									Weighted ***‡		
	1971	1975	1981	1985	1991	1993	1995	1996	1998	1998	2000	2001
Married/cohabiting couple†	2.0	2.0	1.9	1.8	1.9	1.9	1.9	1.9	1.9	1.8	1.8	1.8
Married couple	..	..	..	..	..	..	..	1.9	1.9	1.8	1.8	1.9
Cohabiting couple	..	..	..	..	..	..	..	1.7	1.7	1.7	1.7	1.6
Lone parent	1.8	1.7	1.6	1.6	1.7	1.7	1.7	1.7	1.7	1.6	1.7	1.7
Total: all families with dependent children	2.0	1.9	1.8	1.8	1.8	1.8	1.8	1.8	1.8	1.8	1.8	1.8
*Unweighted sample***												
Married/cohabiting couple	4482	4299	3887	2890	2541	2453	2358	2329	2004		1804	2004
Married couple	..	..	..	..	..	..	..	2086	1753		1558	1720
Cohabiting couple	..	..	..	..	..	..	..	243	251		246	284
Lone parent	382	477	558	458	595	682	658	635	652		660	682
Total	4864	4776	4445	3348	3136	3135	3016	2964	2656		2464	2686

* Dependent children are persons aged under 16, or aged 16-18 and in full-time education, in the family unit, and living in the household.
† Including married women whose husbands were not defined as resident in the household.
* Trend tables show unweighted and weighted figures for 1998 to give an indication of the effect of the weighting. For the weighted data (1998, 2000 and 2001) the weighted base (000's) is the base for percentages. Unweighted data (up to 1998) are based on the unweighted sample.
*** Weighted bases are shown in Table 3.21

FIGURE 13.3 A typical outcome of a data search from the ONS web site

As you can see when you begin to deal with real statistics, they do become more complex. We need to consider carefully the categories of interest (e.g. lone parents) and the importance of definitions and weighting (you would have come across this concept when considering the inclusion of goods in the calculation of the retail prices index).

You can also read or download the 'Guide to Official Statistics' produced by the Office for National Statistics. This details sources of information on population, health, education, the labour markets, housing, crime, household finance, the economy, banking, commerce, transport, agriculture, the environment, civic affairs and government.

Your local library may keep a number of government publications on the shelves. Important collections of statistics are contained in the following (the descriptions are taken from the web site).

The Annual Abstract of Statistics

'This product contains a comprehensive collection of statistics covering the nation.

It contains statistics on the United Kingdom's economy, industry, society and demography presented in easy to read tables and backed up with explanatory notes and definitions. It is compiled from 100 sources and has more than 10 000 series.

It covers the following areas: area; parliamentary elections; overseas aid; defence; population and vital statistics; education; labour market; personal income; expenditure and wealth; health; social protection; crime and justice; lifestyles; environment; water and housing; transport and communications; national accounts; prices; government finance; external trade and investment; research and development; agriculture; fisheries and food; production; banking...'

The Monthly Digest of Statistics

'An important reference work containing the latest monthly and quarterly business, economic and social data.

Twenty chapters of tables, covering the following topics: National accounts, including Gross Domestic Product (GDP); Population and vital statistics; Labour market; Social services; Law enforcement; Agriculture, food, drinks and tobacco; Production, output and costs; Energy; Chemicals; Metals, engineering and vehicles; Textiles and other manufactures; Construction; Transport; Retailing; External trade in goods; UK balance of payments; Government finance; Prices and wages i.e. Retail Prices Index (RPI); Leisure; and Weather.'

Financial Statistics

'Financial Statistics provides key financial and monetary statistics for the UK.

It contains data on public sector finance; central government revenue and expenditure; money supply and credit; banks and building societies; interest and exchange rates; financial accounts; capital issues; balance sheets and balance of payments. From the October 1998 edition the data in Financial Statistics became consistent with the European System of Accounts (ESA 95).

Current data for this publication may be found in the Time Series data facility on a selection basis – see below for link.

Financial Statistics (FS) is a monthly publication. It contains over 200 pages. Part I comprises free-standing data and the figures in Chapters 1 to 7 are updated monthly; part 1 contains data from ONS...'

Economic Trends

'A monthly compendium of statistics and articles on the UK economy, including some regional and international statistics.

Contains data on UK economic accounts; prices; labour market; output and demand indicators; selected financial statistics; GDP; consumer and wholesale price indices; households' final consumption expenditure; final expenditure prices index; visible and invisible trade balance; earnings; and regional and international economic indicators. Includes articles on national accounting; trade; wider economic issues; research and development statistics and international comparisons.

From the October 1998 edition, the national accounts data in Economic Trends is consistent with the European System of Accounts (ESA 95).'

Social Trends

The information provided is shown below in the format you would find on the web site (figure 13.4).

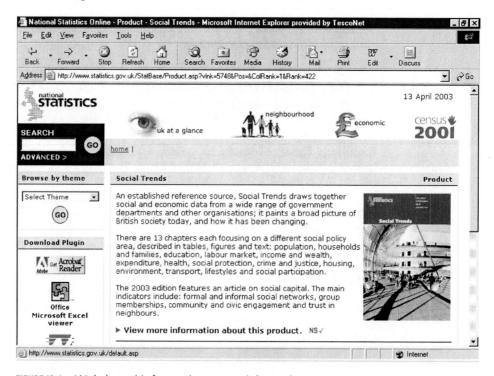

FIGURE 13.4   Web-based information on social trends

13.4 | OTHER SOURCES OF DATA FROM THE WEB

Try some of the following:

http://www.data-archive.ac.uk
This data archive holds over 4000 data sets on social and economic topics

www.bankofengland.co.uk
UK economic reports

www.doh.gov.uk/public/stats1.htm
Give information on the statistics produced by the Department of Health

www.europa.eu.int/comm/eurostat/Public/datashop/ EN?catalogue=Eurostat
Produced by the Statistical Office of the European Communities (Eurostat)

www.london-research.gov.uk
Gives data on London

www.nisra.gov.uk
Range of data for Northern Ireland

www.oecd.org/statsportal
Statistics produced by Organization for Economic Cooperation and Development

www.ons.gov.uk
The web site for the Office of National Statistics. Gives information on the census and range of social surveys

www.un.org/depts/unsd
A range of statistics produced by the United Nations statistical division

The inclusion of these web sites is not intended to indicate any particular importance but rather illustrate the extensive range of web sites now available. There is no shortage of secondary data on the web. The skill is to develop the search strategies that will get you to the valid and up to date data you require, that will be acceptable to others.

You need to bear in mind that internet sites are created for a purpose. If you are researching a controversial topic, you will find sites providing data supporting and opposing the main propositions. You will also find a range of opinion represented. If you doubt this, try searching for 'vivisection' or 'fox hunting'. Data from sites of this type should always be used with care and you need to be aware that your own 'point of view' can cloud your own judgement on emotive issues. The expectation is that the researcher remains objective. However objective, we still work with the data and experience we have.

13.5 | MAKING AN INTERNET SEARCH

Surfing the net can uncover a wide range of data. If you already have a web address, like those given above, you can go directly for information otherwise you will need to use a search engine. Popular search engines include:

Google	**www.google.com** (or **www.google.co.uk**)
Yahoo	**www.yahoo.com** (or **www.yahoo.co.uk**)

Unconstrained, a search engine can give an overwhelming response. Asking Google to search (the UK) for 'secondary data' found 289 000 sites. You need to be careful of your choice of words and aware that your specified search might miss web sites of real interest. If you are interested in the risks associated with scuba diving, for example, you will find very different sites if you search for 'diving incidents' rather than 'diving accidents'.

The search process might allow you to specify connections between key words. Typical connections include + (alternatively ADD) and OR. If we were interested in productivity in the car industry we could link words using 'productivity + car + industry'. If however, we were interested in either cars or vans we could write 'cars OR vans'. You should find the online instructions sufficiently easy to follow.

13.6 | CONCLUSIONS

There is plenty of data out there. Research is rarely done in isolation and it is worth finding out what other work has been done. Using data collected for other purposes (secondary data) should give you good background information and may allow you to develop your own ideas.

You should be able to find helpful data searching the internet. We particularly recommend that you visit **www.statistics.gov.uk**. Having looked at this other data it will be easier to decide whether you need to collect additional information for your own purposes. You should always question the reliability of the data you get. Sophisticated analysis is no substitute for poor data.

Comparing and contrasting your results from primary research (see Chapter 12) to secondary data (this chapter), qualitative research and other sources of information is referred to as *triangulation*. This use of different research methods and techniques in the same study is seen as a way of overcoming possible bias and the limitations of looking at a problem in only one way. Different methods of research allow the cross-checking of results. Any contradictory findings would suggest further work is needed either on the research methods employed, or the understanding of the topic area, or both.

When you have worked through and understood this chapter you will be in a position to:

- identify a number of sources of useful statistical data
- describe the type of information available on a range of business, economic and societal issues
- think more about the search process and the importance of being selective with information

13.7 EXERCISES

1 Suggest a few topic titles and explain how the use of secondary data could be helpful.

2 Identify three other sources of web-based data not included in this chapter.

3 Find a few surprising facts about the UK today.

4 Show how you might use brainstorming to generate ideas regarding holidays in Wales.

5 Show how a 'relevance tree' could help you research a topic like student finance.

6 Find the values of the retail prices index for 'one-person pensioner households'.

7 Browse through some of the sources of information given and suggest at least three project titles.

8 Use a search engine to find one interesting web site. Explain why you find it interesting and comment on its reliability.

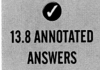

13.8 ANNOTATED ANSWERS

1 It is difficult to imagine any areas of research that cannot benefit from the selective use of secondary data. Even making an informed choice about project title and possible objectives can be facilitated by a knowledge of the work undertaken by others. Whether you are interested in some sector analysis of the economy, or aspect of social change or business characteristics the changes are that statistics exist somewhere. We suggest you try **www.statistics.gov.uk** as a starting point.

Most research will involve some search of the available literature. You could ask at your local library whether packages like ABI Proquest (which also gives access to a number of UK newspapers from the mid 1990s) and Emerald are available.

2 When looking for web-based information, you have lots of choice. Why not try **www.dti.gov.uk**, **www.ft.com** or **www.ukonline.gov.uk**?

3 This question gives you a chance to practise your research skills and explore an area of interest. Surprise us! Have you found the number (estimated) of caravan miles travelled each year, or the highest risk water sport?

4 The topic of interest may be 'holidays in Wales' but this is very broad and you may need to develop your ideas before you give a focus to your enquiry. The chances are that before any research takes place the topic would become more specific and time based.

Brainstorming should allow all kinds of ideas to emerge. Judgement should be deferred. The kinds of ideas you might get:

going up mountains
travelling by cars
days out from Brum
a treat for the kids
old fashioned
on your bike
meals out
you do it with mum and dad
youth hostels
go by coach
and so on

You can cluster ideas of interest. You can again consider your project title and the purpose of your work. Brainstorming should free up your thinking; judging and evaluating ideas is an additional stage.

5 A 'relevance tree' can help you collate your research ideas. In terms of student finance you may be interested in the debt problem now and how students can pay off debt in the future. You could begin to represent your ideas as follows (but there are other ways of developing such a relevance tree):

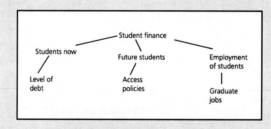

6 If you search **www.statistics.gov.uk** you will find the following extract:

RPI: One person-pensioner households index		
All items (excluding housing)		
Jan 87 100		
2002	Q1	154.7
	Q2	155.3
	Q3	155.0
	Q4	156.1

This extract is for illustration and you can add earlier years. The information is there and you need to be careful about its use. You need to be clear about the definition of a one-person pensioner household and need to be aware of the reasons why housing has been excluded (all in the supplementary notes).

7 This may be a more challenging question than it seems. Many students will say that getting a good project title is the difficult bit. Once you have a focus in your work and a motivational topic, then the rest (your words) is more likely to follow.

Given the information coming to our desks, we have come up with the following project titles:

> 'Why should a relatively small organization generate so many emails?' (you should see our inboxes!)
> 'What skills developed by academic study make a student more employable?'
> 'Do improved numeracy skills improve overall course performance?'

We are sure you will think of lots of your own.

8 We will admit to a bias in our search process and keep finding the web site for *Quantitative Methods: A Short Course*. We believe it supports the book by providing additional material including extra questions. We know that a number of students have found this kind of support useful and we are confident that you will find it easy to access.

WEB REFERENCE 👁

www.thomsonlearning.co.uk/businessandmanagement/curwin3

See the companion web site for further questions and annotated answers. There you will also find a PowerPoint presentation which you can use to help understand this area, or later, for revision. The site also contains links to other sites of interest.

Chapter 14

CONCLUSIONS AND PREPARING FOR ASSESSMENT

Learning something about using numbers to consider and analyze situations is usually one of the objectives of a short course in 'quants'. However, just doing the course is not enough; you are expected to show, in some way, that you have understood the concepts and can use them. The ways in which you might be tested can vary, but there is guidance that might help you succeed. Algebra and formal mathematics are usually tested by some form of examination, but statistics, investment appraisal or linear programming often require the completion of coursework.

We give lots of examples in this chapter of the type of work you might be asked to do (but not the annotated answers). This chapter is about thinking through how you tackle problems. The model answers are, however, given on the web site.

OBJECTIVES

After reading this chapter you should be able to:

- analyze what is required in an assignment
- plan how to succeed in an assignment
- prepare for an exam
- plan how to pass in the exam room

14.1 | COURSEWORK AND ASSIGNMENTS

Most 'quants' courses will use coursework(s) or assignments in their assessment strategies. One reason for this is that it allows the tutor to assess understanding in the context of a 'real world' situation, which can be analyzed in depth, rather than a contrived exercise, which is constrained to take about 45 minutes. Even if you can manage simple exercises, such coursework offers a different challenge since it often requires you to combine ideas from many areas of statistics at once.

The nature of coursework itself has changed so that now the learning objectives and assessment criteria are normally clearly spelled out to you before you start. This is a considerable advantage since you can see exactly what is required to achieve a pass or a mark in the 70s. If you are given coursework which does not include assessment criteria, then try to find what they are. General guidelines are given in your definitive course document (something which goes by many names). If in doubt ask – if nothing else it may remind the tutor to hand them out. (One of us once took an exam where no-one was told the pass mark!)

Having got the question and the assessment criteria you need to *ensure that you understand what needs doing* – this is sometimes less than obvious. Most coursework is designed to take some time, so the option of doing it the night before it is due to be handed in is probably not open to you. Quite often you need to collect data in statistics coursework, and this is likely to be quite time consuming.

Example assignment in Data Collection

You will need to form groups for this exercise. You should aim for a group size of four. You will need to agree any different group size with your tutor.

Design a questionnaire to assess peoples' attitudes to stricter asylum rules that would automatically reject people from 'safe' countries. You should design approximately 15 questions, pilot these with a group of about ten people, and amend as necessary. Propose appropriate quotas for drawing a representative sample of the general public and determine the numbers in each quota for a sample size of 200.

You will need to prepare a report on the work that you have done, including an introduction and conclusion. You will need to submit the questionnaire and the quota sizes worked out by your group. Your report should also include a section from each group member describing their experience of developing and testing the questions, in no more than 500 words each.

Assessment Criteria

Your report should be well structured and include:

> the piloted questionnaire
>
> an assessment of each question
>
> a review of the results of your piloting
>
> suggestions for alterations where necessary and
>
> the quotas and their calculation.

The reflective section should concentrate on the individual experience of asking the questions and the kinds of response given.

70 per cent +: an excellent assignment fully researched and justified
55 per cent to 69 per cent: good arguments and most justifications
40 per cent to 54 per cent: a basic set of results with some personal reflection

Maybe the first lesson here is not to ignore the coursework. Looking at what you need to do shows that there is considerable organization and time management necessary if you are to have anything to hand in. You also need to manage working in a group. Breaking it down as in table 14.1, we have:

1	Find a group to work with	a couple of days
2	Decide who does what	effective group meeting(s)
3	Get some background information on the topic	at least a day
4	Write some questions	at least an hour
5	Discuss them with the group	at least an hour
6	Agree on a final questionnaire	at least an hour
7	Find and interview ten people	a couple of days
8	Amend the questionnaire	at least an hour
9	Calculate quotas	an hour
10	Write 500-word report	two hours

TABLE 14.1   Breaking down coursework into manageable chunks

This might be spread over an eight- to ten-week period, and you will be doing lots of other things too, but it does indicate that you cannot do it over a weekend.

The second consideration is the sort of mark you are trying to achieve. The assessment criteria spell out what you need to do to get certain marks and it is up to you to meet these. In this case, even to get a pass at 40 per cent, you are expected to attempt all of the sections. This implies that if you missed out, for example, the reflective 500-word report, that you would fail. If in doubt, clarify with a tutor whether or not it is possible to pass if you have missed out a section. In general, our advice is to *attempt all sections of any coursework* – you get more marks that way.

It is worth looking at other coursework involving more numbers and some analysis:

Example assignment that requires the collection of data, some description of data and the use of regression analysis

This is an individual assignment.

You have been asked to investigate the way house prices vary in a city that is near to you. It has been suggested that house prices depend primarily on a small number of factors (or variables) including the distance from the centre of the city, the number of bedrooms, a score for house type, a score for garage type and a score for central heating. The score can be allocated as follows:

Score for house type	Score for garage	Score for central heating
1 = terraced	0 = no garage	0 = no central heating
2 = semi-detached	1 = garage	1 = part central heating
3 = detached	2 = double garage	2 = full central heating

You have been advised that you can use the statistical procedure called regression that could help you predict a price given such information. The usefulness of such a procedure will depend on the quality of the data you collect.

You are required to:

1 Collect relevant data for a least 20 houses. You should briefly explain your sampling method
2 Determine a few descriptive statistics and produce two diagrams. Comment on your results.
3 Using Excel, or some other suitable statistical package, produce a plot of house price against each of the other factors. In this case house price is on the *y* axis and the explanatory factor, say distance from the centre of

the city, on the x-axis. Using an appropriate measure, explain briefly how house price relates to each of the possible explanatory factors.

4 To develop a model to predict house price, you have decided to allocate a number of points to each house. This sum of points (X) is to be calculated as follows:

$$X = \text{(miles from the centre of the city)} + (6 \times \text{the number of bedrooms}) + (8 \times \text{score for house type}) + (2 \times \text{score for garage}) + \text{(score for central heating)}$$

Using Excel, or some other suitable statistical package, plot house price, Y, against this sum of points, X. Determine the correlation coefficient and the regression line. Discuss your results.

5 Explain how the model used in part 4 could be improved. Suggest other factors that could be important in the determination of house price.

This coursework involves using the concepts and techniques covered in several parts of the course. You will need to ensure that you are sufficiently familiar with the methods of collecting data, the use of charts and diagrams, the calculation of statistics like the mean, median and mode and the use of regression. In addition, you will need to be able to use a computer package like Excel. You should see this type of assignment as an opportunity to revise and consolidate your understanding of these topic areas. Assessment criteria can be presented in a variety of ways. In this example, we will consider skill in providing evidence.

Assessment Criteria

Your submitted assignment should provide evidence of the:

- ability to effectively present the outcome of a research activity including the collation of data and the analysis of results

- collection of meaningful data including a justification of survey methodology

- selective use of diagrams and descriptive statistics

- use of an appropriate computer package

- correctness of calculations

- correct use of simple regression

This coursework has left the students with a number of choices including:

- should they take a sample of just 20 houses or should they take a larger sample?
- what statistics should they calculate?
- should they use the computer for calculations?
- what computer package should they use?
- what measure should they use for the relationship between house price and other factors?

The answers will depend what the students wish to achieve. The assignment is all about demonstrating understanding. Taking a sample of 40 or 400 or even 4000 houses is not necessarily the way to a good mark (or an easy life!). It is more a matter of justification. If the city of interest has six distinctive areas, for example, it could be argued that at least four houses or five houses or six houses are taken in each area given the resource constraints and for this reason a sample of 24 or 30 or 36 was taken. Make a case for what you do. The chances are that your course will have a preferred statistical package and you are advised to use this. It is likely that you will be expected to use Excel. If you are not familiar with Excel, we would suggest it is worthwhile learning. Given that the assignment has said 'determine' rather than 'calculate' we would suggest you use Excel to produce a range of statistics (such as the average price, the median price, the mode if this is meaningful and the standard deviation) and the two diagrams. You could just produce simple diagrams, such as pie charts showing house type and score for garage. You could also take the opportunity to evidence your skills by producing a more complex diagram that did show the distribution of price and within each price band a key showing area (you would expect extra marks for this).

The assignment does expect the correct determination (calculation) of the correlation coefficient (an appropriate measure of how house price relates other possible explanatory factors) and regression. To usefully discuss possible improvements you should be familiar with multiple regression and the importance of other factors, such as area. There is a lot of work to be done. The assessment criteria provide a useful checklist to ensure that you have at least done the minimum.

14.2 TESTS AND EXAMS

One way of assessing basic maths is to use short tests, often held in seminar/ tutorial slots. These are sometimes used for basic statistics too. A typical example might be:

Section one

1 You have been given a list of 30 students attending a recent seminar on statistics and you are interested in their response to the teaching of statistics on their course. You have been asked to complete the following tasks:

Explain how you would select a random sample of five students (giving all possible respondents an equal chance of selection). (6 marks)

Explain what concerns you might have about using this listing as a sampling frame. (6 marks)

2 The data below gives the number of faults found in a sample of cars after a routine service had been completed:

| 0 | 1 | 2 | 1 | 0 | 0 | 3 | 1 | 3 | 3 | 0 |
| 2 | 2 | 4 | 0 | 3 | 0 | 1 | 2 | 0 | 1 | 0 |

Present this data as a frequency distribution. (6 marks)

3 You have been given the following sample results:

Response to smoking question	Percentage in sample
Smokers	28
Non-smokers	68
No response	4

Present this information using a suitable diagram. (6 marks)

4 You have given the number of hours of overtime recorded in the last week for manual workers in a small business:

Overtime worked last week	Number of employees
under 1 hour	8
1 but under 2 hours	7
2 but under 3 hours	5
3 but under 5 hours	8
5 but under 10 hours	10

Present this information using a suitable diagram. (6 marks)

5 The data below gives the number of faults found in a sample of cars after a routine service had been completed:

0	1	2	1	0	0	3	1	3	3	0
2	2	4	0	3	0	1	2	0	1	0

Determine the mean, (3 marks), median (3 marks), mode (2 marks).

6 You have been given the number of hours of overtime recorded in the last week for manual workers in a small business:

Overtime worked last week	Number of employees
under 1 hour	8
1 but under 2 hours	7
2 but under 3 hours	5
3 but under 5 hours	8
5 but under 10 hours	10

Determine the mean. (12 marks)

Section two

7 $52 - 7 + 3 \times 4 =$ (4 marks)

8 $((2 \times 7 + 3) - 14)/2 =$ (4 marks)

9 Simplify: $10X + 6Y - 7X = 14Y - 6X + 12Y + X$ (4 marks)

10 Simplify: $4X(Y - 2X) - 2Y(2X + 4Y)$ (4 marks)

11 Simplify: $a^3 \times a^5/a^4$ (4 marks)

12 Expand the brackets: $(2x - 5)(2x + 6)$ (4 marks)

13 Find the roots of: $X^2 - X - 12$ (4 marks)

14 Find the roots of: $X^2 - 8X + 15$ (5 marks)

15 Solve the following simultaneous equations:

$$2X + 4Y = 26$$
$$8X - 3Y = 9$$

(5 marks)

16 A linear demand function $(P = f(Q))$ goes through the points:

$$Q = 2, P = 92; \text{ and } Q = 10, P = 60$$

Find the equation of the function. (4 marks)

17 A linear supply function ($P = f(Q)$) goes through the points:

$$Q = 3, P = 18 \text{ and } Q = 15, P = 90$$

Find the equation of the supply function. (4 marks)

18 If a market has the demand function from Question 16 and the supply function from Question 17, find the point of equilibrium. (4 marks)

This test was designed to last an hour and consists of short answers. Students were expected to attempt all questions in both sections. How would you prepare for such a test, and how might you try to maximize your score?

Our suggestions may seem obvious, but provide a simple checklist of things to do (it is not unknown for people to realize that their passport is still at home on arrival at an airport). These suggestions apply to a formal end of semester examination as well as to a mid-term test.

1 Make sure you know when and where the test is being held.
2 Make sure you know what topics are included, and which are excluded from the test.
3 Find out if you can take anything into the test such as calculators, notes, books, etc.
4 Try to get a copy of a previous test paper for your course – often not available.
5 Practice, and then practice again.
6 Ask for help if you run into difficulties during the course or during revision. It is pointless to go to a tutor and just say you need help! If you have attempted some work, got some right but got stuck on other parts, you will receive a much more sympathetic hearing than if you have done nothing.
7 Get to the room on time, or even a few minutes early. Don't panic.
8 Look at the paper and check if you have to do every question, or if you have a choice.
9 Split your time up in relation to the marks. If a question carries half of the marks, spend up to half of the time on it.
10 Whatever happens, attempt the number of questions expected. This maximizes the number of marks you could get. If there were four questions each carrying 25 marks and you only do three of them, the most you can get is 75 per cent.
11 If you get stuck on a question, go on to another and come back later. Leave space to add further work later.
12 Clearly number/label your answers.

A typical end of module or unit examination is now shown. You should:

• *Be aware of time*. This is a two-hour paper with five minutes' reading time. We are becoming more aware of one-hour and one-and-half-hour end of semester examinations and you will need to carefully time manage these.

- *Be aware of structure*. This paper has three sections. The sections marks are not equally weighted. Section A carries 25 per cent of the marks, section B 25 per cent of the marks and section C 50 per cent of the marks. Unless there are good reasons to do otherwise, perhaps you should be thinking about half an hour for each of the first two section and one hour for the final section.
- Section A is testing basic sums. Each question is only worth two and a half marks. Do what you can (we would hope that you could do all of these), and leave space for those you have a problem with or will take too much time.
- Section B requires you to answer only one question from two. You should be able to spot that question 11 is about linear programming and question 12 is about discounting a cashflow.
- Section C requires you to answer two questions from four. You should again be able to spot the four topic areas: basic statistics, index numbers, time series and probability.
- *Be prepared*. You will need to know the calculator that you are going to use. It would also be helpful if you could see the previous examination papers and the formula sheet (not included here). The formula sheet should really help you through certain questions, particularly those where formula are more difficult to remember. It is also useful to know that a table of present values is included.

Name of institution

Type of programme

SEMESTER * EXAMINATIONS ****

SUBJECT:	**Name of module/unit**
SUBJECT CODE:	****
DATE:	**/**/**
TIME ALLOWED:	**2 hours**
READING TIME:	**5 minutes**
EXAMINER:	*****************

INSTRUCTIONS

There are THREE SECTIONS **to this paper**.

Answer ALL **questions in Section A**	**(25% of the marks)**
Answer ONE **question in Section B**	**(25% of the marks)**
Answer TWO **questions in Section C**	**(50% of the marks)**

Calculators may be used but you do so at your own risk.

- graph paper is provided
- a formula sheet is provided (see Formula Sheet on inside back cover)
- a table of Present Value Functions is provided (see Appendix 1)

THIS IS A CLOSED BOOK EXAMINATION

SECTION A

Answer ALL questions in this section by writing in the space provided

	Answer
1 Evaluate: $$19 + 6 + 6 \times 2 - 3$$	
2 Simplify the expression: $$3y(3 - 2x) - 2(3x + xy) - 4x$$	
3 Solve the simultaneous equations: $$6x - y = 22$$ $$2x + 5y = 18$$	$x =$ $y =$
4 If you invest £299 at 5 per cent under a simple interest contract for 3 years, how much interest do you receive?	Interest =
5 If you invest £120 at 4 per cent compound for 3 years, how much interest do you earn?	Interest =
6 Find the roots of the quadratic equation: $$x^2 - x - 6 = 0$$	$x =$
7 A company has a revenue function of $$R = 40Q$$ And a cost function of $$C = 600 + 10Q$$ Find the break-even point	$Q =$ $R =$ $C =$
8 Find the roots of: $$-x^2 - 4x + 17 = 0$$	
9 A linear demand function goes through the points $Q = 5$, $P = 70$ and $Q = 8$ and $P = 40$. Find the equation of the function.	

10   A simple market has
    Demand: $P = 350 - 11SQ$
    Supply: $P = 100 + ½Q$
Find the equilibrium point
	$Q =$
	$P =$

SECTION B

Answer ONE question from this section

11 The Zendo Company has asked you to advise them on the amounts of each of their two products to produce. For the first product, Zendo Original takes five hours of labour time and 30 kg of raw material, whilst the other product, Zendo Deluxe takes four hours of labour time and 100 kg of raw material. Contribution to profit from the Zendo Original is £4 per unit. Contribution to profit from Zendo Deluxe is £7 per unit. Upon investigation, you find that there are 3030 kg of raw material available per week and the labour force works for 220 hours per week. There is an agreement that no more than 30 Zendo Originals will be produced per week.

Formulate this problem.	(15 marks)
Construct a graph and solve the problem.	(5 marks)
Given your solution, advise the company.	(5 marks)

12 The 4STOP company is evaluating various marketing approaches and has reduced the choice to two possible strategies as suggested by two sets of consultants. The strategy suggested by Green's would have an initial cost of £50 000. The strategy suggested by Shah's would cost more initially, at £60 000. The two consultants' projections of revenues are shown in the table below:

	Projected Revenues	
Year	Green's	Shah's
0	£0.00	£0.00
1	£30 000.00	£70 000.00
2	£40 000.00	£80 000.00
3	£50 000.00	£40 000.00
4	£50 000.00	£10 000.00
5	£40 000.00	£0.00

4STOP uses a discount rate of 8 per cent.

a Which consultants marketing strategy should be chosen according to the net present value methodology? (18 marks)

b What reservations might you have about these calculations? (7 marks)

SECTION C

Answer TWO questions from this section.

13 **a** You have been asked to advise on how a sample of 50 students from your college or university could be selected. Describe how such a sample could be selected and the steps you think should be taken to ensure that the sample was representative. (6 marks)

   b You have been given the following data on the number of lectures missed by a group of 30 students:

2	1	0	1	6	0
3	0	2	0	2	2
0	2	1	0	0	0
1	5	1	5	0	4
3	0	3	1	4	0

   i Present this data in the form of a frequency table. (5 marks)

   ii Using the frequency table constructed in part i, present this data as a diagram. (5 marks)

   iii Using either the original data, or the frequency distribution determine the mean, median and mode. (9 marks)

14 **a** You have been given the following sales (in £000s) achieved by a travel company:

Year	1996	1997	1998	1999	2000	2001	2002	2003
Sales	5.4	6.8	7.3	6.9	6.8	7.3	8.6	8.9

   i Using 2000 as the base year (= 100), construct an index showing sales during this period. (5 marks)

   ii Determine the percentage change from 2000 to 2001 and the percentage change from 2000 to 2003. (2 marks)

   iii Determine the percentage change from 1996 to 2000. (3 marks)

   b The following table gives the hourly pay rate to three groups of part-time workers and the number of hours worked by each group in 2001, 2002 and 2003:

| | 2001 | | 2002 | | 2003 | |
Group	Hourly pay rate (£s)	Total hours worked	Hourly pay rate (£s)	Total hours worked	Hourly pay rate (£s)	Total hours worked
Cleaning	3.50	500	4.85	480	4.90	480
Catering	4.80	200	5.80	200	5.90	210
Secretarial	5.10	300	5.90	290	6.00	280

    i   Calculate a Laspeyres index of wages using 2001 as the base year. (5 marks)

    ii   Calculate the Paasche index of wages using 2001 as the base year. (5 marks)

    iii   Comment on your results from parts i and ii. (5 marks)

15 The sales of a particular travel guide have been recorded as follows:

Year	Quarter 1	Quarter 2	Quarter 3	Quarter 4
2001	80	122	130	77
2002	82	118	129	82
2003	83	120	130	

    i   Graph the data and comment on any observed trend or pattern. (8 marks)

    ii   Calculate a centred four-quarter moving average trend and show this on your graph. (7 marks)

    iii   Using an appropriate model, predict sales made through the web site for the 1st and 2nd quarters of 2003. (10 marks)

16 a   A business analyst has classified recent business opportunities as follows:

| | Retail | | Manufacturing | |
Business size	Established business	New business	Established business	New business
Small	85	105	10	18
Large	20	30	62	75

Determine the probability of a business opportunity being in:

i   manufacturing   (3 marks)

ii   a small business   (2 marks)

iii   an established business   (3 marks)

iv   an established business given the additional information that it is manufacturing   (2 marks)

v   an established business given the additional information that it is in large business.   (3 marks)

b   You have been asked to advise a company on the choice between two projects, XX and YY. In both cases there are some uncertainties about the outcomes and these have been expressed in terms of probability, as follows:

Project XX

Probability	Profit (£s)
0.5	5000
0.5	8000

Project YY

Probability	Profit (£s)
0.2	3000
0.3	3500
0.2	4500
0.2	8000
0.1	12 000

i   Determine the expected payoff for each project and on the basis of your results, make a recommendation to the company.   (7 marks)

ii   When making recommendations on the basis of expected value calculations, what would be your main concerns?

          (5 marks)

Remember that we have only given examples of the types of coursework, seminar tests and examinations that are typical. You will need to check the form of assessment used on your course. It is also worth checking whether past papers are available from the library or elsewhere.

14.3 | CONCLUSIONS

This chapter contains little that is new, and many could argue that it is just 'common sense' but our experience suggests that it is helpful to have a study strategy. Quants need not be a difficult subject to study, even if you have not been particularly successful in the past. In many ways it is a matter of confidence. With practice, the methods and approaches will become more familiar. You will begin to know some study areas particularly well. You might be able to draw on other skills, such as the use of Excel. These can all become building blocks. Begin to build the bigger skills like making the most of what you know and being able to learn more effectively. As you begin to build your understanding, assessment becomes an opportunity to show what you can do. *Think success*.

We wish you success (and luck) with your studies.

WEB REFERENCE 👁
www.thomsonlearning.co.uk/businessandmanagement/curwin3

See the companion web site for further questions and annotated answers. There you will also find a PowerPoint presentation which you can use to help understand this area, or later, for revision. The site also contains links to other sites of interest.

Appendix I DISCOUNT FACTORS OR PRESENT VALUE FACTORS

Interest rate	1	2	3	4	5	6	7	8	9	10
Years										
1	0.990099	0.980392	0.970874	0.961538	0.952381	0.943396	0.934579	0.925926	0.917431	0.909091
2	0.980296	0.961169	0.942596	0.924556	0.907029	0.889996	0.873439	0.857339	0.841680	0.826446
3	0.970590	0.942322	0.915142	0.888996	0.863838	0.839619	0.816298	0.793832	0.772183	0.751315
4	0.960980	0.923845	0.888487	0.854804	0.822702	0.792094	0.762895	0.735030	0.708425	0.683013
5	0.951466	0.905731	0.862609	0.821927	0.783526	0.747258	0.712986	0.680583	0.649931	0.620921
6	0.942045	0.887971	0.837484	0.790315	0.746215	0.704961	0.666342	0.630170	0.596267	0.564474
7	0.932718	0.870560	0.813092	0.759918	0.710681	0.665057	0.622750	0.583490	0.547034	0.513158
8	0.923483	0.853490	0.789409	0.730690	0.676839	0.627412	0.582009	0.540269	0.501866	0.466507
9	0.914340	0.836755	0.766417	0.702587	0.644609	0.591898	0.543934	0.500249	0.460428	0.424098
10	0.905287	0.820348	0.744094	0.675564	0.613913	0.558395	0.508349	0.463193	0.422411	0.385543
11	0.896324	0.804263	0.722421	0.649581	0.584679	0.526788	0.475093	0.428883	0.387533	0.350494
12	0.887449	0.788493	0.701380	0.624597	0.556837	0.496969	0.444012	0.397114	0.355535	0.318631
13	0.878663	0.773033	0.680951	0.600574	0.530321	0.468839	0.414964	0.367698	0.326179	0.289664
14	0.869963	0.757875	0.661118	0.577475	0.505068	0.442301	0.387817	0.340461	0.299246	0.263331
15	0.861349	0.743015	0.641862	0.555265	0.481017	0.417265	0.362446	0.315242	0.274538	0.239392
16	0.852821	0.728446	0.623167	0.533908	0.458112	0.393646	0.338735	0.291890	0.251870	0.217629
17	0.844377	0.714163	0.605016	0.513373	0.436297	0.371364	0.316574	0.270269	0.231073	0.197845
18	0.836017	0.700159	0.587395	0.493628	0.415521	0.350344	0.295864	0.250249	0.211994	0.179859
19	0.827740	0.686431	0.570286	0.474642	0.395734	0.330513	0.276508	0.231712	0.194490	0.163508
20	0.819544	0.672971	0.553676	0.456387	0.376889	0.311805	0.258419	0.214548	0.178431	0.148644

Interest rate	11	12	13	14	15	16	17	18	19	20
Years										
1	0.900901	0.892857	0.884956	0.877193	0.869565	0.862069	0.854701	0.847458	0.840336	0.833333
2	0.811622	0.797194	0.783147	0.769468	0.756144	0.743163	0.730514	0.718184	0.706165	0.694444
3	0.731191	0.711780	0.693050	0.674972	0.657516	0.640658	0.624371	0.608631	0.593416	0.578704
4	0.658731	0.635518	0.613319	0.592080	0.571753	0.552291	0.533650	0.515789	0.498669	0.482253
5	0.593451	0.567427	0.542760	0.519369	0.497177	0.476113	0.456111	0.437109	0.419049	0.401878
6	0.534641	0.506631	0.480319	0.455587	0.432328	0.410442	0.389839	0.370432	0.352142	0.334898
7	0.481658	0.452349	0.425061	0.399637	0.375937	0.353830	0.333195	0.313925	0.295918	0.279082
8	0.433926	0.403883	0.376160	0.350559	0.326902	0.305025	0.284782	0.266038	0.248671	0.232568
9	0.390925	0.360610	0.332885	0.307508	0.284262	0.262953	0.243404	0.225456	0.208967	0.193807
10	0.352184	0.321973	0.294588	0.269744	0.247185	0.226684	0.208037	0.191064	0.175602	0.161506
11	0.317283	0.287476	0.260698	0.236617	0.214943	0.195417	0.177810	0.161919	0.147565	0.134588
12	0.285841	0.256675	0.230706	0.207559	0.186907	0.168463	0.151974	0.137220	0.124004	0.112157
13	0.257514	0.229174	0.204165	0.182069	0.162528	0.145227	0.129892	0.116288	0.104205	0.093464
14	0.231995	0.204620	0.180677	0.159710	0.141329	0.125195	0.111019	0.098549	0.087567	0.077887
15	0.209004	0.182696	0.159891	0.140096	0.122894	0.107927	0.094888	0.083516	0.073586	0.064905
16	0.188292	0.163122	0.141496	0.122892	0.106865	0.093041	0.081101	0.070776	0.061837	0.054088
17	0.169633	0.145644	0.125218	0.107800	0.092926	0.080207	0.069317	0.059980	0.051964	0.045073
18	0.152822	0.130040	0.110812	0.094561	0.080805	0.069144	0.059245	0.050830	0.043667	0.037561
19	0.137678	0.116107	0.098064	0.082948	0.070265	0.059607	0.050637	0.043077	0.036695	0.031301
20	0.124034	0.103667	0.086782	0.072762	0.061100	0.051385	0.043280	0.036506	0.030836	0.026084

Appendix II AREAS IN THE RIGHT-HAND TAIL OF THE NORMAL DISTRIBUTION

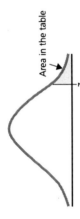

Area in the table

z	.00	.01	.02	.03	.04	.05	.06	.07	.08	.09
0.0	.5000	.4960	.4920	.4880	.4840	.4801	.4761	.4721	.4681	.4641
0.1	.4602	.4562	.4522	.4483	.4443	.4404	.4364	.4325	.4286	.4247
0.2	.4207	.4168	.4129	.4090	.4052	.4013	.3974	.3936	.3897	.3859
0.3	.3821	.3783	.3745	.3707	.3669	.3632	.3594	.3557	.3320	.3483
0.4	.3446	.3409	.3372	.3336	.3300	.3264	.3228	.3192	.3156	.3121
0.5	.3085	.3050	.3015	.2981	.2946	.2912	.2877	.2843	.2810	.2776
0.6	.2743	.2709	.2676	.2643	.2611	.2578	.2546	.2514	.2483	.2451
0.7	.2420	.2389	.2358	.2327	.2296	.2266	.2236	.2206	.2177	.2148
0.8	.2119	.2090	.2061	.2033	.2005	.1977	.1949	.1922	.1894	.1867
0.9	.1841	.1814	.1788	.1762	.1736	.1711	.1685	.1660	.1635	.1611
1.0	.1587	.1562	.1539	.1515	.1492	.1496	.1446	.1423	.1401	.1379
1.1	.1357	.1335	.1314	.1292	.1271	.1251	.1230	.1210	.1190	.1170
1.2	.1151	.1132	.1112	.1093	.1075	.1056	.1038	.1020	.1003	.0985
1.3	.0968	.0951	.0934	.0918	.0901	.0885	.0869	.0853	.0838	.0823
1.4	.0808	.0793	.0778	.0764	.0749	.0735	.0721	.0708	.0694	.0681
1.5	.0668	.0655	.0643	.0630	.0618	.0606	.0594	.0582	.0571	.0559
1.6	.0548	.0537	.0526	.0516	.0505	.0495	.0485	.0475	.0465	.0455
1.7	.0446	.0436	.0427	.0418	.0409	.0401	.0392	.0384	.0375	.0367
1.8	.0359	.0351	.0344	.0336	.0329	.0322	.0314	.0307	.0301	.0294
1.9	.0287	.0281	.0274	.0268	.0262	.0256	.0250	.0244	.0239	.0233

z	.00	.01	.02	.03	.04	.05	.06	.07	.08	.09
2.0	.02275	.02222	.02169	.02118	.02068	.02018	.01970	.01923	.01876	.01831
2.1	.01786	.01743	.01700	.01659	.01618	.01578	.01539	.01500	.01463	.01426
2.2	.01390	.01355	.01321	.01287	.01255	.01222	.01191	.01160	.01130	.01101
2.3	.01072	.01044	.01017	.00990	.00964	.00939	.00914	.00889	.00866	.00842
2.4	.00820	.00798	.00776	.00755	.00734	.00714	.00695	.00676	.00657	.00639
2.5	.00621	.00604	.00587	.00570	.00554	.00539	.00523	.00508	.00494	.00480
2.6	.00466	.00453	.00440	.00427	.00415	.00402	.00391	.00379	.00368	.00357
2.7	.00347	.00336	.00326	.00317	.00307	.00298	.00289	.00280	.00272	.00264
2.8	.00256	.00248	.00240	.00233	.00226	.00219	.00212	.00205	.00199	.00193
2.9	.00187	.00181	.00175	.00169	.00164	.00159	.00154	.00149	.00144	.00139
3.0	.00135									
3.1	.00097									
3.2	.00069									
3.3	.00048									
3.4	.00034									
3.5	.00023									
3.6	.00016									
3.7	.00011									
3.8	.00007									
3.9	.00005									
4.0	.00003									

Additive model *(Ch.9, p.195)*
A model that explains differences by adding component parts $(A = T + S + R)$

Algebra *(Ch.2, p.8)*
Uses letters to represent amounts or quantities which could be money, weight, people or whatever. Algebra allows us to develop formulae, or general statements about relationships between things.

Bar chart *(Ch.4, p.70)*
Chart where bars are equal width and length represents frequency

Base year *(Ch.7, p.132)*
Starting point for an index number, given the value of 100

Base-weighted *(Ch.7, p.138)*
Refers to index series, Laspeyres, where base year weights are used throughout the calculations

BEDMAS *(Ch.2, p.8)*
A mnemonic to help you remember the order of doing things – Brackets, Exponentiation, Division, Multiplication, Addition, Subtraction

Bias *(Ch.12, p.266)*
A systematic source of error

Binomial model *(Ch.11, p.245)*
A discrete probability model where there are 2 outcomes and each trial is independent.

Brainstorming *(Ch.13, p.288)*
Describes a type of technique where judgement is deferred and a range of ideas generated

Break-even *(Ch.3, p.45)*
Making zero profit, where total revenue just equals total cost

Categorical data *(Ch.4, p.71)*
Data which can only be classified or put into groups, for example, eye colour or gender

Census *(Ch.12, p.264)*
The inclusion of all those people or items of interest to us

Chance *(Ch.11, p.235)*
A measurable probability that something happens

Coefficient of determination
(Ch.8, p.166)
Is the squared value of correlation (r^2) and is a measure of the amount of variation in the data that we can explain

Combinations *(Ch.11, p.247)*
In probability, the number of ways of selecting r from n

Component bar chart *(Ch.4, p.73)*
Bar chart where each bar represents two or more sub-divisions of the data

Compound interest *(Ch.10, p.220)*
Money invested gains interest but that interest also gains interest in subsequent years

Constant *(Ch.2, p.11)*
Something which does not change, a horizontal line on a graph

Continuous data *(Ch.4, p.79, Ch.5, p.99)*
Data which can take any value, including fractional values, for example, time taken to do something

Contribution *(Ch.3, p.46)*
The difference between selling price and variable cost of one item – used in determining the break-even level of production

Coordinates *(Ch.2, p.10)*
Reference points on a graph, giving the x and y values of a point

Correlation *(Ch.8, p.159)*
A measure of the relationship between two variables and lies in the range –1 to +1

Cumulative frequency *(Ch.5, p.99)*
The number of items with a given value or less

Data *(Ch.12, p.263, Ch.13, p.285)*
All the facts and figures we have collected

Dependent *(Ch.11, p.240)*
In probability, where the outcome of one event affects the outcome of another

Dependent variable *(Ch.8, p.161)*
The variable we are trying to explain. We are looking to relate the values of the dependent variable to the values of other variables.

Discount factor *(Ch.10, p.221)*
Used in investment appraisal to find current time value of money

Discrete data *(Ch.4, p.79)*
Data which can only take whole number (integer) values, for example, number of children in a family

Equally likely *(Ch.11, p.236)*
Things with equal probabilities

Expected value *(Ch.10, p.226, Ch.11, p.241)*
An amount multiplied by its probability, a bit like an average or what might happen over a long period of time

Exponentiation *(Ch.2, p.9)*
Raising a number, or a letter to a power, which could be a whole number or a fraction or a negative number

Extrapolation *(Ch.8, p.169)*
Making forecasts outside the range of the current data

Feasible area *(Ch.3, p.53)*
Area on a graph which meets criterion – often in relation to an inequality

Fixed cost *(Ch.3, p.46)*
A cost incurred whether or not any production takes place, for example, the rent on buildings.

Frequency *(Ch.2, p.26, Ch.5, p.98)*
The number of times a value occurs, brought together into a frequency table

Frequency definition *(Ch.11, p.236)*
In probability, conducting an experiment to determine a probability

Gradient *(Ch.8, p.167)*
The increase in the y value resulting from a unit increase in the x value (slope of the line)

Histogram *(Ch.4., p.78)*
Chart where bars may be of varying widths and their areas represent the frequencies

Independent events *(Ch.11, p.238)*
In probability, where the outcome of one event does not affect the outcome of another

Independent variable *(Ch.8, p.161)*
The variable being used for explanatory purposes

Index number *(Ch.7, p.131)*
Used to make comparisons back to a base year, can be used to amalgamate many items into one index, most famous is Retail Prices Index

Inequality *(Ch.3, p.52)*
Things that are not equal; often used in linear programming

Information *(Ch.12, p.264)*
Data organised in such a way that it informs the user

Intercept *(Ch.2, p.13, Ch.8, p.167)*
The point where a line cuts the y-axis, in a linear function, the value of 'a' also used in regression

Interpolation *(Ch.8, p.169)*
Making forecasts within the range of the current data

Laspeyres index *(Ch.7, p.139)*
A base weighted index series

Linear *(Ch.3, p.47)*
Straight line

Linear function *(Ch.2, p.11)*
A straight line, as in $y = a + bx$

Linear programming *(Ch.3, p.50)*
A technique to find the optimum allocation of scarce resources between competing uses

Lower quartile *(Ch.6, p.114)*
Gives the value one-quarter of the way through an ordered set of data

Mean *(Ch.5, p.96)*
The mean is calculated by adding the given values together and dividing by the number of values

Median *(Ch.5, p.96)*
The middle value of an ordered list

Mid-point *(Ch.5, p.100)*
The half-way point of a given range

Modal group *(Ch.5, p.103)*
The range of values containing the mode

Mode *(Ch.5, p.97)*
The most frequent value

Moving average *(Ch.9, p.191)*
An average calculated for a specified number of data points that moves forward as new data becomes available

Multiple regression *(Ch.8, p.171)*
When two or more x variables are being used to predict the y values

Multiplicative model *(Ch.9, p.195)*
A model that explains differences by multiplying component elements
$(A = T \times S \times R)$

Mutually exclusive events *(Ch.11, p.237)*
Things that cannot happen at the same time, used in probability

Net present value (NPV) *(Ch.10, p.222)*
Method of finding current value of a projected series of future cash flows to evaluate a project and hence choose between competing projects

Non-sampling error *(Ch.12, p.276)*
Those differences that cannot be explained by the sampling process

Ogive *(Ch.5, p.101)*
A graph of cumulative frequency

Origin *(Ch.2, p.10)*
Represents the point where both X and Y are zero on a graph and is where the two (or more) axes cross

Paasche index *(Ch.7, p.142)*
A current weighted index series

Payback *(Ch.3, p.56, Ch.10, p.218)*
Amount of time it takes to recoup the initial investment

Percentage change *(Ch.2, p.27)*
Looking at the increase from some starting point; they are the basis for index numbers, but are also used in comparing sets of data, especially if the units of measurement in the two sets are different.

Percentiles *(Ch.6, p.116)*
Gives the value at a specified percentage through an ordered set of data

Pictogram *(Ch.4, p.78)*
Diagram where a relevant picture or cartoon is used to represent the size of the data

Pie chart *(Ch.4, p.74)*
A circle divided in proportion to amounts of data

Population *(Ch.12, p.264)*
All the people or items of interest to us

Powers *(Ch.2, p.9)*
How many times the number is multiplied together, also known as exponentiation

Pre-coded answers *(Ch.4, p.71)*
On a questionnaire where the acceptable answers are specified with the question and the respondent chooses one of these

Primary data *(Ch.12, p.264)*
Facts and figures we collect for our purposes

Quadratic function *(Ch.2, p.15, Ch.2, p.47)*
A function where the power of x is 2 giving a graph with a single bend or turning point

Quota sampling *(Ch.12, p.269)*
Selection based on achieving certain numbers with defined characteristics

Range *(Ch.6, p.112)*
The difference between the largest and smallest values

Rank correlation *(Ch.8, p.173)*
Provides a measure of how columns of ranked data relate

Rebasing *(Ch.7, p.135)*
Moving the base year of an index series, often to time align two or more series for comparative purposes.

Regression *(Ch.8, p.167)*
The fitting of a line to a scatter of points

Root *(Ch.2, p.19, Ch.3, p.49)*
The values of x where a function crosses the x-axis – a quadratic has 2 roots, a cubic 3 roots, and so on

Sample *(Ch.12, p.264)*
A selection from all those people or items of interest to us

Sample space *(Ch.11, p.237)*
A diagram showing all of the outcomes in a probability problem

Sampling error *(Ch.12, p.276)*
The variation from sample to sample (part of the sampling process)

Sampling frame *(Ch.12, p.266)*
A listing of all those people or items of interest to us

Sampling unit *(Ch.12, p.264)*
The item, object or person being selected

Sampling with replacement
(Ch.11, p.240)
Where each one selected for the sample from

the population is put back before the next sample member is selected.

Seasonal effects *(Ch.3, p.44, Ch.9, p.195)*
Variations in a time series associated with a particular period of time, say a quarter, or a month, or the differences in the data that can be explained by the season of the year

Secondary data *(Ch.12, p.264)*
Data collected by others that might or might not be of use to us

Simple random sampling *(Ch.12, p.266)*
Selection when all people or items are given an equal chance of inclusion

Simultaneous equations *(Ch.2, p.22)*
Two or more equations which are true at the same time, and usually you need to find the values of x and y where this happens, often needed in solving linear programming problems

Slope *(Ch.2, p.13)*
The value of b in a linear function, also used in regression

Standard deviation *(Ch.6, p.112)*
Provides a measure of the average difference from the mean

Standard normal distribution *(Ch.11, p.249)*
A normal distribution where the mean is zero and the standard deviation is one. Forms the basis for all normal distribution tables of areas.

Statistic *(Ch.5, p.96)*
A descriptive number

Stratified sampling *(Ch.12, p.267)*
Selection using information about the parts/structure of the population

Subjective probability *(Ch.11, p.237)*
A probability found by questioning people, rather than by calculation

Systematic sampling *(Ch.12, p.268)*
Selection using a calculated interval through the sampling list

Time series *(Ch.3, p.44, Ch.9, p.185)*
Data which is measured over time, usually at specific intervals, often represented by a graph, or data specifically collected over time

Total cost *(Ch.3, p.46)*
All costs incurred by the firm, usually seen as fixed cost plus variable cost

Total revenue *(Ch.3, p.46)*
All of the income to a company, in a simple case, just price times the number of items sold

Trend *(Ch.9, p.185)*
General movement in the data over time

Upper quartile *(Ch.6, p.114)*
Gives the value three-quarters of the way through an ordered set of data

Variable cost *(Ch.3, p.46)*
Costs which are incurred as a result of production and which change with the level of production, for example, the cost of raw materials

Variance *(Ch.6, p.113)*
Provides a measure of the average squared difference from the mean

INDEX

EXCEL GUIDE FOR PC USERS

Mark M.H. Goode

Introduction

Excel, a very powerful spreadsheet application produced by the Microsoft Corporation, is now considered to be the industry standard for spreadsheets. Excel is also part of the Microsoft Office range of programs which includes Word (a word processor), PowerPoint (a presentation package) and Access (a database). Computer spreadsheets can store, manipulate, analyze data, draw graphs and charts, as well as provide powerful database operations. Furthermore, frequently performed tasks can be automated by the use of macros. This User Guide has been designed to help you to get started with Excel; it covers all the basic operations and skills.

System requirements

The basic requirements to run Excel are any Pentium PC using Windows 95 or later with 16Mb of Ram. However, although Excel will run under such configuration, a Pentium machine with at least 32Mb of RAM is recommended for more serious work. Excel can also be run on Macintosh computers, although in this Guide we will only be dealing with PCs.

Starting Excel

Excel can only be run from within Windows. To run Excel simply double click the left hand mouse button when the cursor is on the Excel Icon.

Exiting Excel

To exit from Excel, select **Exit** from the **File** menu using the left hand mouse button. You will be asked if you want to save changes that you have made to any documents. Choose the **Yes** button to save changes. Choose the **No** button to quit Excel without saving any changes. Choose the **Cancel** button to return to Excel without quitting.

How to get help

While you are using Excel you can get contextual help at any time by selecting the **Help** menu, **Contents** or press the **F1** key. If you need assistance on using the help facility itself, select the **Help** menu, **How to use Help** within the **Help window**. For some menu commands the software can give graphical demonstrations of how to use or undertake particular operations.

The spreadsheet

The spreadsheet is comprised of three basic areas, namely the worksheet, the menu bar and the tool bar. The worksheet is made up of cells defined by row numbers (these run from 1 to 8192) and column letters (these run from A to IV). There are over two million cells on the spreadsheet, which can store text, numbers or formulae.

Basic moves around the spreadsheet

You can move around the spreadsheet using either the cursor keypad or the mouse. The cursor keypad can be used to move the cursor up, down, left and right using the **arrow keys**. The cursor can also be moved from anywhere on the spreadsheet to cell A1 by pressing the **CTRL** and **Home** keys together. The toolbar and menu bar are activated by using the mouse to point to the area you wish to use and pressing the left hand mouse button.

The menu bar

The menu bar contains all the menus used to draw graphs, open files, save files, change the width of columns, copy and move blocks of information and much more. These menus are structured like a tree with secondary menus off the main menus and tertiary menus off the secondary menus.

The toolbar

The standard toolbar (which is shown above) is displayed across the top of the window below the menu bar to allow quick mouse access via its buttons to the most frequently used commands. This toolbar can be reposition, re-organized or extended by selecting it with the mouse pointer.

Shortcut menus

When you are more familiar with the spreadsheet many commands can be accessed by the short-cut menus when you click the right hand mouse button. The shortcut menus contain the most useful commands for the cell, chart, or other objects that you have selected. A list of the more commonly used short-cut menus, are given in Appendix B at the end of this guide. When you become more familiar with using the spreadsheet the short-cut menus can be used to save time.

Entering data

Three types of data can be entered in any cell on the spreadsheet

1	Text	any string of information proceeded by a ',' or ^. This can be either a text string or number string, e.g. *'Month* or *'50S*
2	Numbers	any number, e.g. *12.89* or *–500.3*
3	Formula	a special function used to undertake calculations, e.g. =SUM(C1..C10) This adds all the numbers in the cells C1 to C10

Basic formulae

One major feature of using a spreadsheet is that basic calculations can be done between any cells which contain numerical data. Once the formula has been entered into a cell the result will automatically update if any of the cells referred to in the formula change.

To add the numbers in cells A1 to A2, and then put the result in A3, move the cursor to cell A3, either by using the cursor keys or by pointing to it with the mouse and clicking the left mouse button once. Then type in the formula which must start with an = sign or a special function.

A
1 29.2
2 33.1
3   Formula in this cell is =A1+A2

Add cells	= A1+A2
Multiply cells	= A1*A2
Subtract cells	= A1–A2
Divide cells	= A1/A2

Advanced or special formulae

Over two hundred advanced formulae are available within Excel that allows the calculation of statistical, financial and mathematical functions (see Appendix A at the end of this Guide). Here is a list of the most common functions:

=AVERAGE(C1..C10)	calculates the arithmetic average
=SUM(C1..C10)	calculates the sum
=MAX(C1..C10)	finds the maximum number
=MIN(C1..C10)	finds the minimum number
=STDEV(C1..C10)	calculates the standard deviation
=SQRT(C1)	calculates the square root

Moving and copying blocks of data

A very useful feature of spreadsheets is the ability to repeat frequently used operations by copying, or to change the appearance of the screen and printed output by moving things around. The process for copying or moving data on spreadsheets is very similar.

Blocks

Not only is it possible to copy or move a single item, or cell, but also it is possible to do the same for a collection of cells, or block. A block is a collection of adjoining cells forming a rectangle on the spreadsheet. This block is identified by the top left and bottom right cells. When written in formulae these two cell references are separated by two dots, for example:

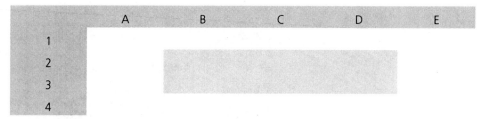

The block above would appear in formulae as B2..D3.

Defining blocks

Blocks can be highlighted using the cursor keys or the mouse.

With cursor keys:

Move to any corner of the block. Hold down the **shift key** and use the cursor keys to move to the opposite corner of the block. As you move the current cell the block will appear in reverse highlighting. Release the **shift key** once the desired block is highlighted.

With the mouse:

The process is similar. Move the mouse to point at one of the corner cells and then press the left mouse button. Keep holding this down while moving the mouse to the opposite corner before releasing it. As with the cursor keys the defined block will appear reverse highlighted. This method of defining a block works best when the whole of the block is displayed on one screen.

Moving a block

A block may be moved or copied using a two-stage process. First the original block is highlighted and then it is either cut or copied from the spreadsheet by selecting the **Cut** or **Copy** option from the edit menu. The **delete key** will also have the same effect as the cut option. If no block is highlighted the action will be performed on the current cell. Once the block is in the clipboard it may be pasted as many times as desired to new locations in the spreadsheet using the **paste icon** or the **Paste** option from the **Edit** menu. The block will appear with the top left of the copied block being placed at the position of the current cell when the pasting was performed.

Special note about copying formulae and fixing cells

Excel will automatically update the referencing of formulae when the copy command is used, incrementing cell references. However, sometimes you do not want this to happen so the anchor or dollar command is used. A $ sign before a letter, in a formula, will fix the column, a $ sign before the number will fix the row. So the formula A2 will fix both the column and row i.e. the cell.

Saving files

Excel will always ask you whether you wish to save any changes when you exit the spreadsheet. There will, however, be other times when you may wish to save the spreadsheet. **It is good practice to save your work frequently and on more than one disk**. This will protect you from computer failures and from errors which you introduce when trying new features or functions. It is always a good idea to save the spreadsheet before attempting to sort data since this process often produces unforeseen results and can be very difficult to undo.

There are many ways of saving your work to a file. The easiest methods are to click on the icon which shows a **floppy disk** or to hold down the control button and type **S** on the keyboard, **Ctrl + S**. Both these methods will automatically save an existing file with the same name. They will, therefore, replace your previously saved version, so check that this new version is what you want before saving it. If you have never saved the file before, Excel will automatically ask you to specify the name you wish to give the file, its type and the drive on which you wish to save it. It is always worth saving your files onto one or more floppy disks, as well as the hard disk of your machine – if you have spent many hours creating a spreadsheet the last thing you want to do is lose it all. To save it, point to the down arrow beside the drives box and highlight the option which says **A:** by clicking the left mouse button while pointing to it. Unless you have a special requirement for one particular file format, ignore the file type option; Excel will automatically select its own. You must, however, give the file a name. Try to make your names meaningful so that you can remember them in the future. You don't need to specify the file extension (the three characters following the full stop) as Excel will insert them automatically depending on the file type. Your Excel files will normally be given the extension **.XLS**.

Occasionally you won't want to replace an existing copy of your spreadsheet but you will want to add a similar copy to disk. In this case you will need to save the spreadsheet with a new name. The process to do this is slightly more complicated than before. Point to the **File** menu with the mouse, single click on it to reveal the options. There is a **save option** which will save the spreadsheet with its current name, as above, but there is also and option, **Save As**. If you choose this option Excel will always ask you for a file name, type and location. Type the new name and then double click the mouse or choose **OK** to save the file. You can also use the keyboard to save the file to a new name as above. Type **Alt + F** together to bring up the file menu. Then press **A** to choose the **Save As** option. Note that you can always select an option on these menus by typing the underlined character in the option description.

Opening files

Previously saved files may be opened and worked on again. Since Excel allows you to work on more than one spreadsheet at once, there are two possibilities. You may retrieve a file, which first closes any current files, or you may open a new file, which retains all current files until you either close them or exit from Excel. If the current file has been altered since it was last saved and you attempt to retrieve a file you will be warned that any changes that you made to the current file will lost if you proceed. You may choose to continue and lose your changes or cancel the retrieve command so that you can first save your current work. To retrieve an existing file, use the **File** menu. You can click on this menu at the top of the screen with the mouse or you may press **Alt + F** together. Choose the **Open** option by clicking the left mouse button while pointing to this option or by typing **O** on the keyboard.

Drawing graphs

Graphs provide a very effective means of presenting information. Excel has powerful charting facilities that allow professional-looking graphs to be produced very quickly and easily. Graphs may be produced in their own windows, separate from the data on which they are based, or they may be inserted into a spreadsheet to support the text and figures around them.

Creating a new graph

A graph can be created in one of two ways, either by selecting the **Insert** option on the **menu bar** and then selecting **chart**, or by using the **chart wizard** speed button icon on the tool bar. The **chart wizard** button looks like a bar chart. Once the chart option has been chosen there are five steps to creating a graph using data on the spreadsheet.

Five steps to draw a graph

*You will need to press the **Next** button with your mouse pointer to move from one step to another.*

1 Enter the range (top left corner and bottom right corner) of the data on the spreadsheet which you wish to graph (e.g. A1..A30).
2 Choose the graph type.
3 Choose the graph format.
4 View the sample graph.
5 Enter any labels or titles you wish to put on your graph.

Some of the most useful types of graphs are listed below:

Bar	Useful for counting frequencies within nominal classes or classes of similar class width.
Line	Useful for data that varies on a continuous scale along both axes and where the x-axis variables are evenly spaced.
X-Y	Useful for regression type relationships between x and y where the x-axis variable is not equally spaced or ordered. Often used for producing scatter plots.
Pie	Useful for showing proportions of a total by nominal categories.
Stacked bar	Similar uses to a pie chart.

These are by no means the full range of options but will be adequate for most purposes. These options may be combined with a 90-degree rotation or displayed in two or three dimensions.

Producing hard copies

The simplest way to produce a hard copy of the spreadsheet is to highlight the part of the spreadsheet that you wish to print and then to select either the **printer** smart icon or to select the **Print** option from the **File** menu. The latter method requires you to type **Alt** + **F** followed by **P**. It is usually advisable to preview prior to printing. While it is not easy to spot small errors in a large print job from the screen it will eliminate the all too frequent simple errors such as not all the data fitting onto a printed page, or the inadvertent omission of column titles and row descriptions. The **Print Preview** option may be chosen immediately prior to printing by clicking the **Print Preview** button with the mouse or by typing **P**. Every page of printing will be available for viewing. Click the left and right arrows in the icon bar at the top of the preview screen with the mouse or use the **Page Up** and **Page Down** keys to move between pages. Use the **plus** and **minus** characters to zoom into or zoom out from the page and make the image larger or smaller. Alternatively point to the previewed page and click the left mouse button to zoom in, or the right mouse button to zoom out. Use the **scroll bars** at the side and below the previewed image, to move around the page. If you are happy with your previewed image and you have access to a printer press the **printer** icon to commence printing.

It is also possible and preferable to choose to print by first choosing to preview the document prior to printing. This can be done by clicking the **Print Preview** icon in the tool bar or by choosing the **Print Preview** option from the **File** menu. To do this type **Alt** + **F**, followed by **V**.

Setting up the page

Access by selecting **Page Setup** from the **File** menu. **Alt** + **F**, followed by **U**.

The report will normally start printing from the top left corner of the page but inset by the size of the margins. The size of these margins can be adjusted using the **Page Set-up** menu. In addition, you may type a brief **Header and Footer** text line, which will be printed on every page. You may choose to centre the print on the page or to reduce oversized reports so that they will fit within the designated pages. Note: Automatic scaling will never enlarge a print to fill a page but will frequently reduce it. If you wish to enlarge the printout, you may choose to scale the print manually by choosing a **Scaling factor** of more than 100 per cent. You may also specify whether you wish to print it in portrait or landscape layout. Portrait is taller than it is wide, like this document, whereas in landscape mode the print is rotated by 90 degrees so that the print is wider than it is tall. You also have a choice depending on the printer chosen for printing to select the paper size.

Print options

These are a further set of options which you may use to enhance the appearance of your printed output. These options may be accessed by pressing the **spreadsheet type** icon from the **Print Preview** screen.

You may wish to print a small section of your data but use headings from elsewhere in the spreadsheet. These may be specified as ranges entered as print option headings. You may choose to place gridlines between all cells, show the row numbers and column letters and even show the formulae used in each cell.

Summary

The basic areas covered in this document will help you get started with Excel, however it should be remembered that there is much more to this programme than covered here. For more information there are a number of excellent books on Excel available in all leading retailers.

Appendix A

Basic Mathematical Formulae

There are a number of basic mathematical operations which can be used on the spreadsheet, as shown below:

^	exponential (raise to the power of)
* /	multiplication and division
+ –	addition and subtraction
( )	brackets

Example

$$= ((B4+B6)/3)^2$$

This formula will add the number in cell B4 to the number in B6; divide the result by 3 and then raise this to the power of 2.

The following mathematical symbols can be used in the conditional **IF** function.

=	equal to
<	less than
>	greater than

Special symbols

$$\$$$

The dollar symbol is used to fix the cell row or column when copying formulae around the spreadsheet. A $ in front of the letter will fixed the column (e.g. $A21), a $ in front of the number will fix the row (e.g. A$21). A $ sign in front of the letter and number will fix the cell (e.g. =A21).

Special Functions

There are over 300 different functions, which can be used on a spreadsheet to perform calculations. The simplest is the **SUM** function which will add up all the cells in the given range so =**SUM(A2..B8)** will add up all the cells in the range A2 to B8. The special functions break down into six major areas.

1   Engineering
2   Financial
3   Database
4   Statistical
5   Mathematical
6   Logical

Users can use the Help system, within the spreadsheet, to find out more about the syntax of required commands and examples of their use. Some of the more common functions are listed below, however, this is not an exhaustive list.

Engineering

Bessel	These functions return values that meet the Bessel equation which has various applications in physics and engineering.
Boolean	These functions handle the application of digital logic which is used in testing, setting of actual bits in a number.
Complex numbers	These functions convert or modify a complex number (where a complex number is the square root of a negative number).
Number conversions	These functions convert values from one number system to another (e.g. base 10 to base 2).

Financial

These functions are used to calculate financial values:

Annuity	Used in special types of investment analysis.
Cash flow	Operates a table to record income and expenditure.
CD	Compute the value of a certificate of deposit.
Depreciation	These functions calculate the depreciation over time.
Investments	These functions calculate the net present value of future cash flows.
Stock	These functions calculate the value of common stock.

Database

These functions are like the statistical functions, except they take three arguments (Block, Column and Criteria) into account:

DMAX	The largest numerical value in the given block.
DMIN	The smallest numerical value in the given block.
DSTD	The standard deviation in the given block.
DSUM	The total of all numerical values in the given block.

EXCEL GUIDE

Statistical

These types of functions break down into two major areas, namely descriptive statistics and inferential statistics:

Descriptive statistics	
AVERAGE	The arithmetic mean of a list of data.
CORREL	Calculate the Pearson's correlation coefficient.
COUNT	The number of observations in a data list.
INTERCEPT	Returns the intercept of a regression line.
KURT	The kurtosis (peakedness or flatness) of a data list.
MEDIAN	The median of a data set.
MODE	The mode of a data set.
SLOPE	Returns the slope of a regression line.
STDEV	The standard deviation of a data set.
VAR	The variance of a data set.

Inferential statistics	
CONFIDENCE	Returns the confidence intervals for a population mean.
FDIST	Returns the F distribution function.
NORMSDIST	Calculates the cumulative probability value for the normal distribution.
	Example =normsdist(z) where z is number of standard deviations you want the cumulative normal distribution for.
	Example =normsdist(1.3333) is equal to 0.9087
TDIST	Returns the area under the 'student t' distribution.

Mathematical

ABS	Returns the absolute number (i.e. ignores negative signs).
ACOS	Returns the angle whose cosine is a given number.
ASIN	Returns the angle whose sine is a given number.
ATAN	Returns the angle whose tangent is a given number.
CEILING	Round a number up to the nearest integer.
COS	Returns the cosine of a given angle.
FLOOR	Round a number down to the nearest integer.
INT	Returns the integer part of a number (no rounding).
LOG	Returns the log to the base 10 of a number.
ROUND	Returns a number rounded to the desired level.
SIN	Returns the sine of a given angle.
SQRT	Returns the square root of a given number.
TAN	Returns the tangent of a given angle.

Logical

IF   This evaluates a condition, and returns either true or false based on the stated condition.

Example

= if(logical-test, value_if_true, value_if_false)

The following **if** test looks at the value in cell B3, if it is greater than 4 the comment *'Cell is greater than 4'* is printed, if it is not the comment *'Cell is less than 4'* is printed.

=if(B3>4, 'Cell is greater than 4', 'Cell is less than 4')

If statements can be nested to include other **if** statements to allow for more than two outcomes (e.g. true or false).

FILEEXISTS	Checks to see whether a filename exists or not. Returns true(1) if the file exists or false(0) if it doesn't.
ISERR	Returns 1 if a given cell contains the ERR (error indicator), otherwise 0 is returned.

Appendix B

Short-Cut Menus

This is a list of the more commonly used short-cut commands which can be accessed when you are using the spreadsheet. Some of these short-cut menus are also available in other Microsoft products such as Word for Windows, Access and PowerPoint.

1	Help	F1
2	Edit a cell	F2
3	Repeat the last action	F4
4	Spelling	F7
5	Create new file	Ctrl+N
6	Open file	Ctrl+O
7	Save file	Ctrl+S
8	Print file	Ctrl+P
9	Cut	Ctrl+X
10	Copy	Ctrl+C
11	Paste	Ctrl+V
12	Find	Ctrl+F
13	Replace	Ctrl+H
14	Goto	Ctrl+G

FORMULA SHEET

CHAPTER 2   Roots of a quadratic $y = ax^2 + bx + c$ are at $\dfrac{-b \pm \sqrt{(b^2 - 4ac)}}{2a}$

CHAPTER 5   Mean:

$$\bar{x} = \frac{\sum x}{n}$$

$$\bar{x} = \frac{\sum fx}{n}$$

Median:

$$= l + i\left(\frac{n/2 - F}{f}\right)$$

where l is the lower boundary of the median group

         i is the width of the median group

         F is the cumulative frequency up to the median group

and f is the frequency in the median group

CHAPTER 6   Variance:

$$\frac{\sum (x - \bar{x})^2}{n}$$

$$\frac{\sum f(x - \bar{x})^2}{n}$$

Standard deviation:

$$\sqrt{\frac{\sum (x - \bar{x})^2}{n}}$$

$$\sqrt{\frac{\sum f(x - \bar{x})^2}{n}}$$

$$\sqrt{\left[\frac{\sum fx^2}{n} - \left(\frac{\sum fx}{n}\right)^2\right]}$$

CHAPTER 7   Simple index number: $\{P_n/P_1\} \times 100$

Laspeyres Price Index Year 2 $= \dfrac{\sum (P_2 Q_1)}{\sum (P_1 Q_1)} \times 100$

Laspeyres Quantity Index Year 2 $= \dfrac{\sum (P_1 Q_2)}{\sum (P_1 Q_1)} \times 100$

Paasche Price Index Year 2 $= \dfrac{\sum (P_2 Q_2)}{\sum (P_1 Q_2)} \times 100$

Paasche Quantity Index Year 2 $= \dfrac{\sum (P_2 Q_2)}{\sum (P_2 Q_1)} \times 100$

Value Index Year 2 $= \dfrac{\sum (P_2 Q_2)}{\sum (P_1 Q_1)} \times 100$

CHAPTER 8 Linear correlation coefficient:

$$r = \dfrac{n \sum xy - \sum x \sum y}{\sqrt{\left(n \sum x^2 - \left(\sum x \right)^2 \right)\left(n \sum y^2 - \left(\sum y \right)^2 \right)}}$$

Linear regression for $y = a + bx$

$$b = \dfrac{n \sum xy - \sum x \sum y}{n \sum x^2 - \left(\sum x \right)^2}$$

$$a = \dfrac{\sum y}{n} - b \dfrac{\sum x}{n}$$

Spearman's coefficient of rank correlation: $\quad r = 1 - \dfrac{6 \times \sum d^2}{n(n^2 - 1)}$

Where $\quad n$ is the number of paired observations

and $\quad d$ is the difference of ranks

CHAPTER 10 Compound interest formula: $\quad A_0(1 + r)^n$

Present value $= A_n/(1 + r)^n$

CHAPTER 11 Probability rules

> Rule 1: Where A and B are mutually exclusive, then:
> $$P(A \text{ or } B) = P(A) + P(B)$$
> Rule 2: Where A and B are not mutually exclusive, then:
> $$P(A \text{ or } B) = P(A) + P(B) - P(A \text{ and } B)$$
> Rule 3: Where A and B are independent, then:
> $$P(A \ \& \ B) = P(A) \times P(B)$$

Combinations: $\qquad \dbinom{n}{r} = \dfrac{n!}{r!(n-r)!}$

Binomial: $\qquad P(r \text{ successes in } n \text{ trials}) = \dbinom{n}{r} p^r (1 - p)^{n-r}$

z-scores: $\qquad \dfrac{\text{Value} - \text{Mean}}{\text{Standard deviation}} = \dfrac{X - \mu}{\sigma}$

Lightning Source UK Ltd.
Milton Keynes UK
UKOW040001071011

179845UK00002B/108/P